WILLIAM WARDELL

BUILDING WITH CONVICTION

A. G. Evans

GRACEWING

Originally published in 2010
by Connor Court Publishing Pty Ltd, Australia

This edition 2011

Gracewing
2 Southern Avenue
Leominster
Herefordshire HR6 0QF

ISBN 978 085244 767 3

Cover design by Ian James
Typesetting by Openbook/Howden

"We build not for this generation only, nor for the next, but for those who will exist in centuries far removed from us"

William Wilkinson Wardell

"Architecture is the barometer of Faith ... the revival or decline of true ecclesiastical architecture is commensurate with that of the true Faith."

A. W.N. Pugin

Wardell's Gothic-style signature which he used to sign all his earlier architectural work. He ceased to use the style shortly after he started practice in Australia, preferring a simplified form.

CONTENTS

SYDNEY: 1878-1899

INTRODUCTION

When the German poet, Heinrich Heine, stood admiring the glorious portals of Amiens Cathedral early in the nineteenth century, his companion turned to him and asked, 'why could we no longer build such piles?' Heine himself tells us he replied: 'Dear Alphonse, men in those days had convictions, we moderns have opinions. It requires something more than an opinion to build a Gothic cathedral.'[1]

That 'conviction' of which Heine spoke is intrinsic to both St Mary's Cathedral in Sydney and St Patrick's in Melbourne. Conviction makes them such successful buildings; both are acknowledged worldwide to be the finest examples of neo-Gothic design of the nineteenth century. Their architect, William Wilkinson Wardell, was a convert to Catholicism whose work throughout his life was grounded in his faith; a man imbued with Catholic convictions, not mere opinions.

It is given to few architects to design cathedrals; those who have done so in the past two hundred years in Australia (both Anglican and Catholic) may be fewer than half a dozen. Some, like William Butterfield and A.W. Pugin, designed from afar, never setting foot on Australian soil.

William Wardell, however, emigrated from England with his family and settled in Melbourne in 1858. He had already established a busy, successful practice in London and had built some thirty churches in the Gothic style across England and Scotland. He arrived in Australia with a reputation already firmly established, and soon became a dominant figure in his profession in Melbourne at that time. He had known John Henry (later Cardinal) Newman, and had been an acquaintance and admirer of the leading Catholic architect and designer, A. W. Pugin, both of whom exercised a strong influence on him – Newman on his spiritual development, and Pugin on his architecture. Cardinal Wiseman praised Wardell's work, and Henry Manning (later to become Cardinal and to succeed Wiseman) expressed his sorrow at bidding goodbye to a dear friend, writing: 'your absence will be a subject of grief to all who know you, and a loss to the Catholic body.'[2]

In addition to designing these two cathedrals, Wardell was also responsible for some of the grandest and most valued public buildings in Victoria and New South Wales. Probably the best known of these, and instantly recognisable, is Government House in Melbourne.

Although he designed in different styles, it is for his Gothic Revival work that Wardell will be best remembered. His early travels studying the cathedrals of England and France had convinced him of the superiority of Gothic for ecclesiastical buildings and in this he was a disciple of Pugin and the other Gothic Revival apologists. Pugin collaborated on some of Wardell's English church interiors, notably Our Ladye Star of the Sea, Greenwich, and he seems to have maintained an influence on Wardell throughout his career. The philosophies of both – and even their lives - converged at various times.

We cannot ignore cathedrals. From earliest times they have occupied the most prominent places in cities. In the ages of Catholic Faith the lesser buildings of the cities were grouped around and below them as if in humble obeisance. Mediaeval cathedrals were designed as a celebration of belief; they were centres of worship, of learning, of social activity and of civic pride. To some extent this is still the case today even in our secular world. Both St Patrick's in Melbourne and St Mary's in Sydney are assets in the civic landscape of both cities. They have become symbols of those cities rivalling Flinders Street Station and trams in Melbourne, and the Harbour Bridge and Opera House in Sydney. The Wardell cathedrals maintain their dominant position, surrounded as they are by towering commercial blocks – those spiritless cathedrals of commerce - and are, in a sense, a visual refuge from them.

The majority of the populace cares about what the cathedrals look like from the outside, and also cares about their surroundings. Both St Mary's and St. Patrick's benefit from their garden settings. The Cathedrals' familiarity and townscape are somehow comforting and inspiring. Thousands of sightseers, attracted by the grandeur of the cathedrals, venture inside to marvel at their height, the rows of soaring pillars, the aisles and the clerestories above, the sense of space and of length, and the jewel-casket effect of the liberal use of stained glass. Inquisitive non-believers are invariably drawn inside cathedrals whose otherworldly grandeur acts as a mysterious magnet. Few visitors are untouched by the power and significance of their design and wonder at the faith that inspired it. To this extent, if for no other, the casual visitor resembles Winston Churchill who famously admitted that while he was hardly a pillar of the Church of England, he was 'in the nature of a buttress because I support the church from the outside.'[3]

The architect of Australia's two major cathedrals was the antithesis of a mere buttress in the Churchillian sense. He was totally committed to the Faith which his cathedrals enshrined. He believed, as did his mentor, Pugin, that good ecclesiastical buildings could be produced only by architects sharing the Christian Faith. Like Pugin, Wardell saw the Gothic style as embodying and proclaiming in stone and glass the truths of Catholicism, and he believed passionately in the superiority of Gothic over other forms for ecclesiastical buildings. Wardell's ecclesiastical work echoed Pugin who wrote that Gothic design and church furnishings 'embody in their elements and their symbolism the Faith and moral values of Christianity.'[4]

Few if any cathedrals throughout the centuries can be said to have been designed and completed by one architect and in one style throughout. Among English cathedrals, Salisbury may claim to be so, for it was 'built in a single campaign according to the master plan of one directing intelligence.'[5] The main construction which we see and admire today, famously painted by Constable and boasting the tallest spire in England, was completed within thirty-eight years.

Norwich, too, comes close to this ideal - the nave, choir and transepts were completed under the patronage of Bishop Herbert Losinga and consecrated within the space of nineteen years. True, the spires of both cathedrals were added at a much later date long after the original architect had died. This is the case with both St Mary's and St Patrick's Cathedrals in Australia. Although Wardell died before either of his two great Gothic Revival buildings was completed, both were designed, the building processes were supervised, and many of the furnishings were chosen, by the one architect. This gives them a satisfying homogeneity and pleasing conformity that is generally missing from the older cathedrals. In many cases mediaeval cathedrals were altered and added to over centuries in accordance with changing tastes and often at the behest of the local bishop – or perhaps the vanity of the king, as in the case of Westminster Abbey. Much of the interior of Westminster is English Decorated of the thirteenth century and contrasts with the Henry V11 chapel with its elaborate pendant vault added in the fifteenth century in the Perpendicular style. Conversely it can be argued that what is missing from the cathedrals of the Gothic Revival - but present in many of the older foundations - is the subtle mingling of these successive styles and the historical perspective that we gain from exposure to the passing centuries. That sense of history present in the ancient cathedrals we tend to lose in their Gothic Revival equivalents. Nevertheless the doctrinal instruction is still there - the truths of the Faith – which are preserved in the stones and glass as ever they were in the mediaeval cathedrals; and now invariably lost in late twentieth century and post-modern churches.

Thoughtful visitors to St Mary's and St Patrick's Cathedrals who come to admire the architecture, must confront the religious beliefs embodied in the design. As a result the most receptive among them find themselves suspending disbelief, at least for the duration of the visit. As John Betjeman wittily observed of a particular old English country church where 'the beauty makes you gasp' and 'the silent peace brings you to your knees.'[6] Wardell's Cathedrals surely, (like Betjeman's Cornish church), seem to go on praying, even when there is no one in them.[7] Cathedrals, whether mediaeval or Gothic Revival, exercise a special hold over our imaginations; they draw us back and back again. We experience the harmony and majesty in the structure. For the faithful they are doubly uplifting because, in Pugin's words, being 'surrounded by all the holy splendour ... the august mysteries appear ten times more overpowering and majestic'.[8]

William Wardell's name still lives in his Cathedrals although to the general public – even to those who worship in the Cathedrals - he is scarcely known. His grave at Gore Hill cemetery in North Sydney is overgrown and neglected, yet his importance as a major contributor to Australia's architectural landscape cannot be exaggerated. He was also a fascinating character, a direct descendent of that influential band of nineteenth century Gothic Revivalists whose status and influence fluctuated with the vagaries of fashion. In the mid-twentieth century when modernism started to dominate city landscapes, their work was sometimes derided as romantic eclecticism. At the end of the eighteenth century Gothic Revival became a popular fashion, but it is the nature of fashion to swing like a pendulum - to quote Kenneth Clark's metaphor. In the introduction to his influential work 'The Gothic Revival', he states how he, along with other aesthetes of the 1920s, felt themselves 'justified in thinking that these styles are devoid of merit', and notes that the Gothic Revival at that time had barely received a paragraph in a learned periodical.[9] Clark, in the revised edition thirty years later, repudiated these narrow views.

In more recent times when city buildings are built more by accountants than by architects, the Gothic Revivalists have been appreciated anew and their principles reassessed. One of Australia's most distinguished literary figures of the early twentieth century, A. G. Stephens, leading critic, *Bulletin* editor, writer and poet, described Wardell as 'by far the most eminent architect who has lived in Australia';[10] and more recently D. I. McDonald noted at the conclusion of his ADB biographical entry that 'his cathedrals and churches, notable for the purity of expression and richness of symbolism, rank among the greatest buildings constructed anywhere in that style.'[11] Elsewhere McDonald judges Wardell, along with Reed and Blackett, as one of 'the three giants of Australian architecture.'

Few architects ever achieve the degree of fame and public admiration that some painters, playwrights or writers do. But they too are artists whose work, unlike pictures, books and plays, cannot be ignored. As Chesterton has pointed out, you can tear a poem to pieces if you don't like it, or turn a picture to the wall, but you cannot do either to a large public building. 'Architecture approaches nearer than any other art to being irrevocable because it is so difficult to get rid of.'[12] Perhaps if architects were required to be identified publicly with their buildings, extolled or denounced as other artists are, or their work subjected to regular published criticism, the public might be able to exercise more influence on what is placed before them. But architects remain anonymous for the most part, their work unsigned; a tradition that goes back to mediaeval times. Seldom can we blame an individual architect for his mistakes, or praise him for his successes. There are few exceptions. Informed Sydney residents who are asked to name significant architects of their city might get as far as identifying Francis Greenway and Jørn Utzon and perhaps the much-publicised late Harry Seidler, but might have difficulty naming any more.

The purpose of this book is to make the name of William Wilkinson Wardell more widely known and his work better appreciated. It may be argued that there are others of his contemporaries – Edmund Blackett, Joseph Reed, William Tappin, Horbury Hunt among them - who deserve similar recognition. But they are beyond the scope of this work.[13] Wardell's two monumental cathedrals place him squarely in the forefront of the Australian architecture story. Because of these we want to know more about the man who created them, in much the same way that we are inquisitive about the lives of famous composers and writers. Opinions may differ as to who was the greatest Australian architect of his era but there can be no argument that he, more than any other, built from religious convictions and his Catholic Faith imbued his work with an integrity to a degree that - in the biblical phrase - 'the stones cry out'. While the post-modern buildings around the Cathedrals are impermanent and may well be replaced as taste and economic demands swing with the pendulum, the message of the Wardell stones is permanent. They endorse John Ruskin's much-quoted admonition, 'When we build, let us think we build for ever. Let it not be for present delight, nor for present use alone; let it be such a work as our descendents will thank us for.'[14]

Those who come to this biography looking for a definitive record of Wardell's complete architectural work with detailed descriptions and analyses, may be disappointed. I have not attempted a technical manual; this is primarily a biography. The man and his life are placed centre stage, and while little remains of his private correspondence, and for reasons to be explained later his early life has been deliberately obfuscated, much can be learned of him from studying his work and observing his dealings with colleagues. Wardell's architecture is a dominant part of his life, and the reason why his life interests us in the first place. It has therefore received what I hope is adequate coverage, so that the life and the architecture become interdependent. I confess to being unashamedly a Wardell admirer, and I trust that my personal responses to his architecture liberally stated, will be shared by most, if perhaps not by all my readers.
A. G. Evans
April 2010

ENGLAND

1823 – 1858

CHAPTER 1

BIRTH AND EARLY LIFE

William Wilkinson Wardell was born on 27 September, 1823 in the London Borough of Poplar. At age five months, on 3 March the following year, he was carried by his parents, Thomas and Mary Elizabeth Wardell, to the nearby Parish Church of All Saints to be baptised by the Reverend William West, according to the rites of the Church of England.

The terrace house where William was born, number 60 Cotton Street, has long since disappeared. The once narrow street, now greatly widened, is a busy double carriageway being a main route for heavy traffic bound for Canary Wharf, Thames-side industries and docklands further east. The Church, which is a handsome classical-style building with a lofty western steeple and Grecian portico, was then only recently built and consecrated. It was built largely from money given by the East India Company - that dominant source of power, employment and commerce in the Poplar district at that time. The church remains well-used even today, standing in a square of welcome green once the graveyard. Now the old grave mounds are flattened, many anonymous under the grassed area and car park. The rescued, moss-grown headstones lean against the perimeter railing, their inscriptions now so weather-worn that the dead, including William's parents, remain at peace, frustrating any attempt at identification.

The place of William's birth may not be judged particularly auspicious for someone who, in later life, was to become a leading member of his profession and a much admired architect in both England and Australia. Eventually he would become an intimate friend of church leaders in both countries, and be remembered posthumously as a major figure in Australia's architectural history. Fortunately, history often surprises us with stories of pre-eminent lives that grew from relatively impoverished beginnings, and William Wardell's life may be counted among their number.

In the nineteenth century Poplar was (and remains today) part of the East End - as famous in its way as the West End of London. It used to be said that the West End had the money and the East End had the dirt.[1] The Borough falls within the great area of London commonly associated with a poor and densely crowded population, and has generally had an evil reputation among those who don't live there. Now the area attracts a large immigrant population with its cheaper public housing and employment opportunities. At the turn of the century, prior to Wardell's birth, Poplar was scarcely more than a village. But by 1823 the area was developing rapidly following an influx of population attracted by the building of the West and East India Docks and it had lost forever its semi-rural environment. William's birth-place was quickly transformed into a district of narrow, squalid streets and mean houses bounded by Stepney and Limehouse to the west, by Hackney to the north, by the River Lea with its burgeoning manufacturing and chemical industry in the east, and in the south by the River Thames. Here the river snakes its way to the sea making a sudden, southerly U-shaped loop enclosing an area known as the Isle of Dogs. In shape Poplar resembles a narrow rectangle standing on end from south to north. Today, the Local Authority is called Tower Hamlets.[2]

Cutting across the loop formed by the Thames south of Poplar High Street, were sited the East India, West India and London Docks built in the first ten years of the nineteenth century. In Wardell's childhood these docks were crammed with tall ships so that they resembled a forest of masts. Long before industrialisation the forest had been poplar trees from which the Borough is said to have derived its name. Much later, in our time, the docks have given way to a forest of glass and concrete, marking business and residential towers known as Canary Wharf: an affluent area which suggests that the West End no longer has all the money and the East has been cleansed of some of the dirt.

The years immediately following Wardell's birth were to see more frenzied dock building east of London Bridge, resulting in newer docks to cope with the thousands of merchant sailing ships plying back and forth between London and the fabulous East, from 'Distant Orphir', as John Masefield's poem tells us, laden

With a cargo of ivory,
And apes and peacocks,
Sandalwood, cedarwood, and sweet white wine.[3]

Wine was certainly a 'cargo', stored in the seven acres of wine vaults built for the purpose close to the docks and destined for the rich and powerful. Among other more basic commodities would have been spices, coal, tea, sugar and grain.

It will be shown later that the proximity of wine vaults, shipping and civil engineering so close to Wardell's childhood home had a direct bearing on later key events in his life.

If the place of Wardell's birth was inauspicious, the date was hardly less so. He was born in the reign of King William IV who lived on – 'dropsical, drunken, and stupid'[4] – until Wardell was fourteen. The great social and political reforms that were hard-fought for in Parliament and the press, and were enacted in the second half of the century, were yet to be introduced by the reformers; and even afterwards, conditions were slow to improve. Unread by the young Wardell perhaps, the reformist journal *The Edinburgh Review* led the way in exposing what it termed 'the major evils' of the time - the injustices that existed in Wardell's childhood and were to be gradually remedied by parliamentary acts before mid-century. Among those eventually repealed were the Catholic penal laws; the cruel Corn Laws; the Corporate Test Acts; the oppressive Game Laws (man-traps and spring guns were set all over the countryside); prisoners on trial for their lives denied defence counsel; libel punished by the most brutal and vindictive imprisonments; the laws of debt and of conspiracy requiring barbarous penalties; the slave trade still tolerated and Parliament supporting the interests of individual landowners while large sections of the public went unrepresented.[5]

Nothing arouses the sympathy of the modern reader more, and is more likely to expose the Victorians to condemnation, than the condition of the children of the poor in the nineteenth century. Anyone with even a mild interest in Victorian social history would have seen pictures and read about emaciated waifs typified by Jo the Crossing Sweeper in *Bleak House*, or the eponymous *Oliver Twist*. In these examples Dickens was not merely a comic writer; he was a social crusader. Many children were little more than slaves and a scandalous number were child prostitutes. A contemporary observer writing in mid-century estimated that there were some one hundred thousand pauper children in London although how he arrived at this figure is not explained.[6] Young children of the destitute could be 'bought' for as little as a few shillings, or, if able-bodied, for a few pounds. Children, if they worked at all, were often sent underground or worked as chimney-sweeps known popularly as 'climbing boys' of which there were a great number when coal fires were the chief source of heating. Those who had no work and no means of support rummaged in refuse bins or begged.

The cruel and much feared Press Gangs still operated in the Thames area whenever His Majesty's ships needed crew. Pitiable tales are told of young men, some little more than children, being carried off, taken from their wives, their families or their employers. Most of them were totally unsuited to a life at sea and many died there. The plight of Lizzie Hexam and her father combing the river for debris and waste and - most prized of all - the bodies of the drowned as described by Charles Dickens in the opening chapter of *Our Mutual Friend* - was far from mere sensationalised fiction. It is cause for wonder, and was to Australia's eventual gain, that young William avoided the hazards and temptations of the East End streets of those times, and was also spared the mortal epidemics that carried off a high proportion of young children. In 1837, for example, cholera broke out on the hospital ship *Dreadnought* and quickly spread to the surrounding Limehouse district. In 1839 almost half the funerals in London were of children under the age of ten. But by that date Wardell had safely reached the age of sixteen years.

Although shielded from extreme poverty and want, the young William had ample opportunity to observe the pitiable social conditions of the district at first hand. For most of his childhood he lived with his family in Poplar's imposing but grim Workhouse fronting the High Street, not as an impoverished inmate it must be said, but as the son of the Master and Mistress of the establishment. Wardell never forgot his experiences there, nor despised those less fortunate than himself who were incarcerated because of cruel penury. Writing years later in defence of Gothic architecture he is clearly relying on experience when he argues that the features of a Gothic church are a comfort to the poor;

'[Gothic] does not perpetuate to them the hated forms of their workhouse hall or stifling factory; but recall to their minds tales they have heard of those days in England when poverty was not thought a sin, but to be poor was to be relieved, when to be sick was to be visited, when to be hungry was to be fed; and all this, too, not in the modern, grudging spirit of a union workhouse and griping guardians, but with the open-handed unsparing liberality of a Catholic monastery.'[7]

He must have witnessed daily the 'grudging spirit' of the 'griping guardians', the 'hungry not fed', and the 'sick not visited'. Later he mentions the 'contract funerals' and the fear among the poor of ending up on the dissecting table of a local hospital. His verdict on his parents who were in charge of the workhouse – whether they were kind to the inmates, or indifferent to their suffering – is not known. We have no way of knowing whether on this occasion he was writing of his own workhouse, or merely the system in general, but it is clear that his memory of workhouse conditions sickened him.

William's father, Thomas Wardell, is a shadowy figure whose social standing remains uncertain. Described by one writer as a 'small-trader of varying fortune'[8] he is listed as a baker. A later family descendant, perhaps wishing to exaggerate his status, described him as a banker. Whatever the nature of his 'banking' or 'baking', he had at one period, business premises adjoining his house in Cotton Street. This must have been reasonably successful because a Deed of Sale exists to show that he went into partnership with a shipwright, William Lewis, and together they bought land in Wells Street. Evidently the enterprise, whatever it was, failed, because by 1830, when William was seven years old, Thomas Wardell was driven to seek full-time employment at the local Workhouse.

The Trustees of the Parish of All Saints at that time exercised direct control over the Workhouse and would do so for a further six years until a new Act of Parliament came into force. At the Trustees' meeting on 11 November, 1830 they resolved to appoint a new Master and Mistress. It was noted that the previous Master, Thomas Howard, had died and 'the great age and infirmity of his widow renders it expedient that she be retired upon a salary.'[9]

At that same meeting a Committee was appointed to examine the several candidates who had already applied for the positions and it is interesting to note that Thomas Wardell's name was not among them. Evidently he was a late applicant, but his long-standing in the Parish – he was most likely a minor church functionary – would favour him. There were several Wardells in the district, some closely and others distantly related to Thomas the 'baker'. In some documents the name is spelt Wardle as it is in the Trustees' Minutes but William throughout his life, and his family, always wrote Wardell. The name, variously spelt Wardel, Wardill, Wardale, originates on the Durham-Cumberland border and is said to derive from the River Wear in Weardale.

William's mother, Mary Elizabeth, was a Dalton before marriage and her father, also William (after whom the young William was surely named) was a successful Thames Barge owner in business at New Wharf behind Old Palace Yard, Westminster. He was one of two who stood surety for Thomas and his wife - £150 each – which was required of the Master and Mistress of the Workhouse. The other guarantor was John Fuller, a barge builder of Millwall, doubtless a well-to-do colleague of John Dalton's. It was well understood, both by applicants and the Trustees, that the Master of a large workhouse, being responsible for the day-to-day management of funds and contracts, would have opportunities for corrupt practices; surety was a form of guarantee of good behaviour.

The selection process of the new Master and Mistress of the Workhouse took place at the meeting of the Trustees held on December 23 and seems to have been a collusive affair. When a motion was put that the several candidates be called in separately in order to be questioned by those present, it was overruled. Evidently the outcome had already been decided by the late applicant's supporters. A ballot was taken and Thomas Wardle [sic] and his wife Mary Elizabeth, were duly elected by sixty-nine votes - twenty votes ahead of the next candidate in a list of eight. The salary of the Master was stated as £100 per annum, 'without any other emoluments or allowances', and the Mistress (later termed the Matron), £50.[10] While certainly not affluent, the Wardells would have found themselves reasonably well-off if we remember that their accommodation and food would have been provided. They were permitted to appoint servants from amongst the inmates. Those previously trained in domestic work were preferred but if unavailable, the fittest and most presentable were chosen. According to Henry Mayhew, the cost of living was low by our standards, and although the Wardells would not have had to buy food for themselves, prices of these commodities at the time are a useful barometer: butter and tea were five pence per pound, meat one shilling for three pounds, and herrings three or four for a penny. Education fees, when the boys were of school age, could be had for £4 per year rising to £12 for a good grammar school.

In January 1831, when William was seven years old, he and his family moved into the staff quarters of the Workhouse. Their new home was situated in the right wing of the entrance block which also housed the Town Hall where the Trustees met. The handsome building designed in the classical style by James Walker, fronted the south side of the High Street, giving little hint of the grim reality behind it. Hidden from the passing traffic by the grand frontage was the Workhouse proper, a sprawling complex of ramshackle buildings – dormitories, dining hall, kitchens, laundry and chapel – a misery world of 'hated forms of the workhouse hall' as described later by Wardell himself.

William's younger brother, Herbert Samuel, was born in the Workhouse later that same year; his sister, Mary Ann, appeared in 1837, and his younger brother, James Ranford, in 1839.

The position of Master and Mistress was no sinecure. The Master's performance in his job was closely monitored. Regulations and Rules for the Management of the Workhouse were to be strictly followed and the preliminary instructions to the Master show a remarkably benign but high standard to be followed:

'The Master should consider himself placed in a position of great trust and conscientiously discharge every part of his duty with fidelity and honour. As a servant of the Parish he must be frugal in housekeeping, exact in his accounts, humane and attentive to the sick, and strict and positive in his commands; in short he must be a faithful servant, a good Master, and a father to those poor children committed to his care; he must watch every action and not let any crime go unnoticed or unpunished. The strictest justice and impartiality must in all case be administered, and his law and precepts be enforced by his own example. He must be attentive to all goods and provisions delivered into the House and superintend the weighing and the quality thereof and compare the quantity with the Bills of Parcels and regularly keep the accounts for the inspection of the Trustees.'[11]

The document continues with an intimidating list of rules - the keeping of day books, the listing of admissions and discharges, illnesses, births and deaths, and any other incidents. The Master had to superintend the work of the able-bodied. At Poplar, in-house work is listed as Stone Breaking, Oakham Picking, Shot Making, and Pumping Water; and for the women, 'Knitting and Needlework'. There was also appointed a resident schoolmaster paid £36 per annum, and a schoolmistress paid £35. When the Wardells were appointed to the Workhouse there were three-hundred and sixteen inmates made up of one-hundred and thirty-eight males and one-hundred and seventy-eight females.

Our popular image of a Workhouse Master and Matron is influenced forever by Charles Dickens's Mr Bumble and Mrs Corney in *Oliver Twist*. The famous Cruikshank sketch of Oliver asking for more porridge in front of an enraged Mr Bumble has become part of our cultural inheritance so that we have difficulty believing that not all Masters were like Mr Bumble.[12] We should bear in mind, however, that Dickens's grim portrayal of Oliver in the Workhouse first appeared in serial form when William Wardell was fourteen. This post-dates the 1834 Poor Law Amendment Act which removed responsibility for workhouse administration from the relatively humane parish trustee committees and placed it in the hands of Union committees. Poor Law Commissioners were appointed under the new Act, and Unions representing several neighbouring districts were formed. 'Guardians', as they were called, were elected as administrators responsible to the Commissioners. Prior to this date it is generally agreed among historians that the administration of poor relief was more flexible, even if - as was the case in Poplar - the Parish Trustees were hard pressed to reduce rising expenditure. But in spite of their difficulties, there is much evidence that the Trustees of All Saints parish approached their work in a spirit of Christian charity. They objected strongly to the new system being introduced. They disliked being merged with other river-side parishes and, in the Minutes of their Meeting following the publication of the Act, noted that it was,

'a measure opposed to the genuine principles of Christianity in as much as it regards Poverty as a Crime, whereas the divine founder of our Holy Religion was himself Poor...it is a measure vicious in principle having only political economy for its Basis.'[13]

The Poplar Trustees managed to delay action until 1836, but reluctantly had then to adopt the new system. The Poplar Poor Law Union was formed, overseen by an elected Board of Guardians fifteen in number, each representing different districts. The new regulations required that the poor rate be collected locally but removed from parish control only to be administered by the Guardians. They in turn were responsible to the government-funded Poor Law Commission. The Commissioners formulated strict regulations and procedures, collected statistics and produced reports to Parliament. The policy dictated that all relief must be channelled through the Workhouse system, and the procedure required that those seeking relief should be rigorously assessed and were entitled only to very basic and often inadequate sustenance if and when admitted to the Workhouse. These measures were clearly designed to deter the indigent and work-shy poor from seeking relief. The Commissioners saw as their duty the need to make life in the Workhouses less attractive than work for wages elsewhere. They framed regulations to achieve this end; inmates were segregated, wives from husbands, children from parents. Such institutions became hated generally, and were viewed by the poor with horror. The 'poorhouse option' was to be avoided at all cost as one of Dickens's most pathetic characters, the ill and dying Betty Higden, wails: 'I've fought against the Parish [the Workhouse] and fled from it all my life, and I want to die free of it.'[14] Inmates were virtual prisoners and had to seek permission to make short visits outside the walls. Under the new Act Thomas Wardell's managerial powers were circumscribed; he was required to apply to the Guardians before incurring even the most modest expenditure. When an inmate named Farwell applied to be discharged and given temporary 'out relief', Wardell had to seek permission from the Guardians to pay him, and was instructed that two shillings and sixpence be allowed. But when Sara Stephenson applied to be discharged and be allowed money to enable her and her two children to proceed to Sheffield, 'out relief' was deferred and in this instance Wardell was ordered to report back with further details.[15] Outdoor relief was supposedly abolished by the 1834 Act but in many places parish relief and independent charities survived.

The Wardells must have performed their duties to the satisfaction of the Trustees because they were reappointed unopposed by the Union and their salaries actually raised. Whether their lives and routine work of the Workhouse changed much under the new system is not entirely clear. The duties of the Master and Mistress remained much the same, although the more formal regulations came ultimately from a Government Commission and not from the more sympathetic members of their own parish. The Master was adjured to keep accounts, pay other members of the staff and to 'enforce industry, order, punctuality and cleanliness.' He was to see that 'the able-bodied were put to work, to call the medical officer in cases of illness. He had no further concern for the welfare of the inmates except to read prayers morning and evening, and to see that grace was said before and after meals.'[16] The Matron was to take charge of the female paupers and manage the domestic work of the house.

The status, ability and honesty of Masters undoubtedly varied from one workhouse to another. In the larger Unions like Poplar, the Master would have had a clerk to run his office, and pauper servants.

The Wardells would have led a reasonably comfortable existence enhanced further when their joint salary was raised to £180 per year. In addition, the Master was entitled to claim a percentage of the profit derived from the productive labour of the inmates – a policy designed, no doubt, to encourage increased commercial productivity and, in turn, to reduce the claim made upon the public purse. As an example, the Minutes of the Committee for General Purposes on 20 April, 1835 noted that Mr Thomas Wardell was entitled to the sum of £32.14s.0 being ten per-cent of the income derived from the work carried out by the paupers.[17] With sums like these, and doubtless other perks on top of their combined salaries, the Wardells if not affluent, could be considered reasonably well-off.

However, in spite of the domestic comforts and relative security of their new life in the Workhouse, the Wardells would be tainted to some degree by the evil reputation which the system has earned in history and in legend. William Wardell, in later life, could not have escaped entirely the opprobrium of having been so closely connected with the hated establishment. It is perhaps understandable that there is no mention of the Workhouse connection in the few short written accounts of his life, nor in his writings, save the one oblique reference already quoted. Both Wardell himself, and his family descendents, were silent - even evasive - on the matter of William's early life and his relationship with his parents.[18]

Frustrating as this is for the biographer, it is nevertheless understandable in an architect and engineer trying to make his way in the world, and at a later date, establish himself in a new colony in an era over-sensitive on the subject of a settler's origins.

William's boyhood - if not entirely forgotten - was best never alluded to.

Wardell at 27 years old. Sketch in pencil made by his neighbour and close friend, the artist, Clarkson Stanfield R. A.

CHAPTER 2

FROM CHILDHOOD TO MATURITY

William had just passed his eleventh birthday when, in the late afternoon of 16 October, 1834, a great fire broke out in the basement of the House of Lords. Within hours a mighty conflagration had arisen and destroyed the old Houses of Parliament at Westminster.

If he had happened to be at home in Poplar that day he could hardly have been unaware of what was happening five miles up river. A pall of 'white yet dazzling smoke' and at the same time 'a shower of fiery particles'[1] was carried by the westerly wind eastwards and the red glow was visible for many miles around the city. The architect Charles Barry who, as a result of the fire would be forever famous in the history books, was travelling north from Brighton at the time and saw the glow in the sky long before he arrived to witness the incident. Thousands of sightseers lined the banks of the Thames on the Surrey side or 'swarmed upon the bridges, the wharfs and even the housetops ... the spectacle was one of surpassing though terrific splendour.'[2] An eye-witness reported that many thousands were attracted to vantage points to marvel at the stately appearance of Westminster Abbey, 'whose architectural beauties had never been seen to greater advantage than when lighted by the flames of the unfortunate fire.'[3] Fortunately the Abbey was untouched by the flames.

In the crowd on the south bank, close to where the Festival Hall now stands, was J.M.W. Turner sketching the scene and noting the colours for two, later famous, oil paintings. Known for minute observation and the accuracy of his colours, Turner's *The Burning of the Houses of Lords and Commons* provides documentary evidence of the destructive power of the blaze – the sheets of yellow and bright red flames, the smoke rising to a great height spreading and carried in the wind down river - the whole horrific scene reflected in the waters of the Thames.[4]

When light dawned the following day the Houses of Lords and Commons, including the Library and many irreplaceable treasures, were seen to be destroyed; the whole site, other than historic Westminster Hall, was gutted.

The young William, as he listened to the rumours and counter rumours that were brought down river by lighter-men as far as Poplar and beyond, could not have understood the significance of the event historically. Neither could he have guessed how the destruction of the Houses of Parliament and the events that followed would have consequences in shaping his own life and career. A. N. Wilson in his magisterial study, The Victorians, describes it as symbolic, a pivotal event in the process of modernising the British Political system.[5]

It may be over-stating the case to assert that what became known as the Gothic Revival was born with the decision to design a new House of Commons and Lords - the most important public building in England - in the Gothic rather than the Classical style. Some support for this view is given by Kenneth Clark in whose judgement it was, 'a central point in the early history of the Revival'; a symbolic, public anointing of Gothic which led almost immediately afterwards to its popularity.[6] The event and its consequences brought to prominence Wardell's associate and mentor, the young Pugin, whose major contribution to the design of the Houses of Parliament has only recently been fully explored and acknowledged.[7] William Wardell's embrace of the Gothic style was still several years in the future, but clearly he would have been influenced and encouraged by the new popularity of Gothic stemming from the aftermath of the Westminster fire.

Documentary evidence of Wardell's early life and education is meagre. Biographical notes and obituaries are invariably drawn from a sparse memoir by Wardell's grandson, Vincent Andrew Wardell, written in 1940, forty years after his grandfather's death, and a century after Wardell's boyhood.[8] For reasons suggested earlier there was reluctance on the part of family members to talk or write about William's early life in Poplar. The Memoir is confined to generalities, short on facts and dates - perhaps deliberately so. Where William went to school is not mentioned.

William's youngest brother, James Ranford Wardell, boarded at the Collegiate Academy in Horn Lane, Linton, a village south of Cambridge. The self-styled 'superior school' run by John Dorrington and his sister, Mary, was noted for its Classical, Mathematical and Commercial studies. While it is tempting to think that William would have gone there, preceding his younger brother, it is highly unlikely because from the records it seems the school did not open until William had reached the age of fifteen. It seems more likely that he attended a Grammar school, or one of similar standing, and, like his brother's school, well clear of London. Most Victorian parents living in the metropolis at that time and who could afford to do so, sent their children to schools in the country to escape the insanitary and disease-ridden conditions especially prevalent in the densely-populated areas close to the River Thames. The vile odours from the river were so distressing that on one occasion they forced Queen Victoria and Prince Albert to abandon their intended pleasure cruise in 1858. Earlier, Mayhew reported that he gazed in horror as he and his companion saw 'drains and sewers emptying their filthy contents into [the river]; we saw a whole tier of door-less privies built over it [and] heard bucket after bucket of filth splash into it'[9] The threat of typhus and cholera epidemics terrified everyone, rich and poor alike. In addition to the cholera outbreak on the *Dreadnought* moored very close to the Wardell's workhouse, another more serious outbreak swept through London in 1832 when William was nine years old. On that occasion some eight hundred persons died in the East End. Thousands were to die horrible deaths in subsequent outbreaks, a total estimated to number ten thousand in London alone.[10] The desire to get children of school age out of London would account for the popularity of small academies owned and run by Oxford and Cambridge graduates or clergymen; usually with their wives acting as matrons. That William Wardell may have attended one of these schools is conjecture, but highly probable. The alternative was attending a poor local school with the danger of contracting cholera or one of the other contagious diseases – surely unlikely given his father's modest but adequate income. Either way, neither Wardell nor his family makes any reference to his schooling in later documents or anecdotage. Vincent Wardell in his memoir merely states, evasively, that 'it was intended that he should be an engineer,' and was educated with this in mind.[11]

Where he went to school may be of only minimal importance. What seems certain is that he was blessed with precocious intelligence and was capable of continuing his education himself. There is much evidence in his writings that he was well-schooled in Latin, French, and English grammar, and one of his extant note books containing his engineering formulae is testimony to his proficiency in mathematics. He was also a gifted artist and on the evidence of the sketches and designs he has left us, and the opinion of his friends, he could have made art his career.[12]

The proximity of the Thames and the docks crowded with tall ships discharging cargo and leaving again for exotic ports, must have aroused the imagination of boys in the neighbourhood and young William would not have been immune to their attraction. Vincent Wardell tells us – maddeningly without adding further detail - that his grandfather served an apprenticeship at sea when he left school.[13] The late Theresa Wardell, his granddaughter and one-time custodian of family information, and who was generally reliable if purposely discreet, stated that her grandfather 'as a boy ran away to sea'[14] - a proposition that raises questions concerning his relationship with his parents and his attitude to life lived in premises attached to a Workhouse. Running away to sea is a bold, romantic thing for a boy to do and if true, surely would have become a favourite family legend repeated elsewhere. But there is no further evidence and doubt must remain.

There is no record of a William Wardell's name in the Royal, or Merchant Navy lists for the most likely years although there is a George Wardell, aged fourteen in 1838, shown indentured to William Bennett, Master of the *Surprise*, and then again on the *Aladdin*, in 1842. If Teresa Wardell is right about William 'running away to sea' he could have used the name George as a disguise. Another explanation suggests that George was William's cousin and the two boys encouraged each other in spirit of adventure. If 'George' signed on the *Aladdin* in 1842, 'George' could not have been William (see below). Thus we are left in uncertainty about this carefully veiled period. But what is more certain, again according to Teresa Wardell, is that William went to sea against his father's wishes.

Evidently William's enthusiasm for a maritime career did not survive much beyond his first voyage because he asked his father to buy his release from his indentures. This Thomas Wardell declined to do and insisted his son continue until he had completed his contract.[15] A natural enough punishment, we may think, for running away to sea against a father's wishes originally. But if the late Teresa Wardell's reminiscences are to be relied on, sickness intervened and a bout of Yellow Fever led to William's early discharge.

The next we hear of him is his indenture to a Mr Morris, Engineer in private practice and Surveyor to the Commissioners of the London sewers. There he made a thorough study of engineering. His notebook, already mentioned, is packed neatly with trigonometry and other mathematical formulae, dating from the period around 1839 or 1840 when he was seventeen. In 1841 we know from the census that William, aged eighteen, had returned to live at home – home in this case being the staff quarters of Poplar Workhouse. The family at that time consisted of the two parents and three children, William Wilkinson, Herbert Samuel, and James Ranford Wardell. The ages given on the census form were notoriously inaccurate. (William's age is given as fifteen which is clearly wrong given his birth date). Subsequently he entered the office of a London architect, William F. East, probably in 1842. East's office was at 6 Corporation Row, Clerkenwell, a short daily journey for William living in Poplar.[16]

We know little about East except that he was engaged on work connected with railway building at the height of the Boom period when a great patchwork of lines was built or being planned to criss-cross England linking Bristol, Manchester, Leeds, York, Darlington and many other centres. The title 'Architect' was used more loosely than today and was popularly regarded as a branch of engineering. There was no formal training for architects at that period, nor a registration system. As Rosemary Hill notes 'The profession had yet to organise and formalize. Many aspiring architects did train in architectural offices, but it was not a requirement.'[17] A history of the profession states that it was usual for London architects to accept pupils and assistants. Their status was that of articled clerks 'unless they had sufficient status to rank as improvers in which case they would receive a small weekly wage'.[18] They were also expected to attend lectures at the Academy. Undoubtedly there were a good many incompetents or dilettantes in the profession - one thinks of Mr Pecksniff in Dickens' *Martin Chuzzlewit*. Those whose names are remembered and celebrated in the story of architecture - including Wardell - had artistic talent, historical knowledge, and usually a grasp of engineering principles.

We may wonder how William's parents afforded the fees and maintenance of their son, first indentured in the navy, and then in an engineer's office and subsequently with an architect where, if he earned anything at all, it would have amounted to very little. Vincent Wardell mentions 'a rich inheritance from a family friend'[19] and we long to know more. Who this mysterious benefactor was is impossible to name with certainty. One strong possibility is Samuel Wilkinson, one of the Guardians – each Wardell son included second names, Samuel, and Wilkinson, suggesting a strong bond between the families. The presence of this influential Guardian might also explain how Thomas Wardell was voted Master of the Workhouse in the first place. Another possibility is that William's grandfather on his mother's side, William Dalton the prosperous barge builder, would have helped to pay for his grandson's education.

But as we shall see in the next chapter William's 'Great Expectations' unlike those of the young Pip Pirrip, were ceremoniously terminated when he converted to the Catholic Church.

CHAPTER 3

INVENI QUOD QUAESIVI

When the youthful William Wardell came to engineering at some time around the beginning of the 1840s he could hardly have made a more promising choice. Side by side with dock building on the Thames which Wardell would have witnessed daily, these first years of Queen Victoria's reign were marked by an explosion of railway building, popularly dubbed Railway Mania. Proficient engineers and engineer-architects were a favoured species much as IT technicians are today. By 1845 over six hundred railway schemes representing a capital outlay of £563 million had been laid before Parliament. 'Here the whole world is railway mad' wrote Brunel to a friend in France, and added, 'The dreadful scramble ... is by no means a good sample of the way in which work ought to be done.'[1]

By the middle of that decade over two thousand route miles of railway had been laid and the pace of expansion was increasing five-fold.

Where railways opened up, the engineer surveyor had gone before plotting the best routes, and it was this work that Wardell first found himself engaged in. Triangulation and taking astral sights involved many techniques familiar in marine navigation which, we assume, Wardell had studied while at sea. With his skill in mathematics and his application and single-mindedness, he must have established himself as a valued and reliable field officer.

We cannot know at this distance which railway routes he plotted, but it seems Wardell visited several of the English cathedral cities – perhaps big railway centres like York, Durham, Exeter or Norwich, and while in these places found opportunities for studying cathedral architecture. He made drawings of specific features, and laboriously took accurate measurements, amassing a compendium of detail long before many of the cathedrals had been restored, and before the profusion of guide books and detailed studies was published and sold to tourists. It seems inconceivable to us that at the beginning of the nineteenth century the mediaeval cathedrals of Britain and France were neglected disgracefully, in a bad state of decay in many instances, and largely disfavoured in comparison with the classical style. Scarcely one cathedral had escaped mutilation as a result of the Protestant Reformation in England, and the Revolution in France.[2] To the Reformers, cathedrals were uncomfortable reminders of the Old Faith and many were vandalised for this reason. It should not be forgotten that the description 'Gothick' originated as a term of abuse. But with the coming of the Gothic Revivalists and the antiquaries - scholars who studied Gothic buildings 'stone by stone, climbed spires, dug trenches, and dredged lakes'[3] - public perception of Gothic was about to change.

By studying Gothic cathedrals intimately 'stone by stone', Wardell was treading the same path that Pugin had taken a few years before him. It is unlikely that he went as far as Pugin in having himself lowered down a castle well to see whether there was treasure buried there; or exploring a trench which he urged workmen to dig around monastery ruins minutes before it collapsed, the better to understand the mediaeval drainage system. Wardell had a less impetuous nature than Pugin and was given to speculation rather than direct action.

But like Pugin's, Wardell's study of mediaeval cathedrals and their history drew him into the embrace of the Catholic Church. As Pugin had written, 'I learned the truths of the Catholic Religion in the crypts of the old Cathedrals of Europe'.[4] Although speaking figuratively it is clear that it was the history revealed in those crypts, and in what lay above them, that proved a catalyst leading to his conversion. Wardell would have arrived at Catholicism by the same route. 'we look on these remains', he wrote,

> 'with an affectionate and filial devotion knowing that they are the evidence of the glory of our spotless mother ... We visit our old churches and our imagination re-peoples them; they speak to us of our Faith and our Worship and we feel at home in them.'[5]

It is worth remarking that the path to Rome of another young architect with strong Australian connections was directed in a similar fashion by his study of English Cathedrals. John Hawes, the architect-priest who came to Western Australia at a much later period than Wardell (1915) wrote of his excitement when, as a boy at school, he came into contact with the history and pageantry of Canterbury Cathedral. With an artist's eye, bookishness, and a sense of history he 'drank of the cup of tradition' and spent long hours in the Cathedral and in the library. Hawes's path led him to Catholicism via High Church Anglicanism, but Pugin and Wardell opted for Rome without the intermediate High Anglican stepping-stone.[6]

Wardell's path to Catholicism may well have been illuminated by his study of Cathedrals, but the remnants of his library dating from that time show that his conversion was an intellectual one. He read and admired the works of the influential Catholic historian-priest, John Lingard, and possessed his ten-volume history of England, as well as Lingard's *History and Antiquities of the Anglo-Saxon Church*.[7]

The few short accounts of Wardell's life, including an entry in the Australian Dictionary of Biography, refer to Newman's friendship with Wardell and the influence it had on Wardell's decision to enter the Catholic Church. Influenced to some extent by Newman's writings Wardell may have been, but evidence of a close friendship is more tenuous. Neither can Newman's conversion be taken as the final spur that impelled Wardell into the Church. Wardell's baptism at the old chapel of St Mary and St Joseph, Poplar, is dated 5 June 1844, fully a year prior to Newman's discreet reception at Littlemore. We know that Newman was at pains to avoid publicity leading up to his final break with Anglicanism and so it would not have been widely known until a later date. Wardell would have been aware, as all newspaper readers would have been at the time, that Newman had resigned as Vicar of the Oxford University Church of St Mary's and preached his last sermon there on Sunday 24 September, 1843, But only a few intimates knew of this in advance, notably Henry Wilberforce and Pusey who were said to be in tears on that occasion, knowing that their University friend and colleague of so many years had taken his first public step towards an inevitable change of religion. But still Newman hesitated, and remained a member of the established church. He wrote to his bishop, 'I am not relaxing my zeal until it has been disowned by her [the Church's] rulers. I have not retired from her service until I have lost or forfeited her confidence.'[8] On this evidence Wardell could not have been motivated by Newman's conversion – except in a very general sense by the famous Tracts of the Times. Newman's conversion occurred later than Wardell's.

If we are looking for influence it is more likely to have been a combination of several factors: the spirited Catholic Revival in England following the Emancipation Acts of 1829, the Tracts of the Oxford Movement, Wardell's reading of Lingard, his friendship with Dr Daniel Rock, a passionate advocate of Gothic architecture and finally, Pugin's conversion which was publicly known about several years earlier and was regularly defended in Pugin's own writings.

As to the common belief noted by students of Wardell that Newman was an intimate friend of the architect, the evidence again appears tenuous, resting as it does on one extant letter from Newman thanking Wardell for his support during 'the late proceedings against me'.[9] The 'proceedings' relates to the notorious trial when Dr Giacinto Achilli, a defrocked Dominican friar, sued Newman for libel. Achilli had been imprisoned in Rome by order of the Inquisition and after his release toured England as a popular lecturer denouncing the Church and feeding the appetite of the 'No-popery' party avid for lurid scandals and corruption among the clergy in high places. But Achilli's heel was his own disgraceful record of profligacy and hypocrisy, and this Newman alluded to in a series of lectures he gave at the Corn Exchange in London in 1851.

The proceedings dragged on for two years and caused Newman intense distress and anxiety. He was sustained by the prayers and practical efforts of fair-minded friends, one of whom journeyed to Italy to secure witnesses to Achilli's debauchery. The jury found for the plaintiff and Newman was fined £100 – a nominal amount widely considered to indicate that Newman had been vindicated. Even *The Times* was driven to describe the result of the trial as 'unsatisfactory and little calculated to increase the respect of the people for the administration of justice.'[10]

Wardell was evidently one of many who wrote offering his support and Newman's letter in reply was, perhaps, similar to others written to supporters.

'Your very kind letter is one of those many touching demonstrations which have been an abundant and overflowing reward for any pain or anxiety I may have had in the late proceedings against me. I thank you heartily and my kind friends at Liège for their liberality. I hope you will convey my acknowledgement, and inform them, if it is worth while to say it when the matter is so clear already, that I heard the other day in London from two sources that Achilli had arrived in the United States. The London papers, as far as I know, have kept silent on the subject.

Should anything bring you into this neighbourhood I hope you will give me the opportunity of returning to you my thanks in person. [Signed] John H. Newman.[11]

There is no evidence that such a meeting took place in Edgbaston, or anywhere else.

It is difficult for us at this distance to fully understand the religious turbulence of those times. The complacency of the Established Church was being questioned by reform movements, notably the Tracterians led by Newman who dreamed of a 'purified continuum of mediaeval Catholicism.' Catholics themselves, so recently restored to full suffrage following the Emancipation Acts which abolished the penal laws in force since the Reformation, watched these events closely in the expectation that the split in the Anglican Church would result in the re-conversion of England and full union with Rome. Pugin and his circle especially, believed that England was about to experience a Catholic rebirth – Newman famously described the times in a sermon in 1852 as 'The Second Spring,' the first being the coming of St Augustine and conversion of England in the seventh century.

Not all shared the certainty of Pugin, but there certainly existed a palpable optimism among Catholics, and a scarcely suppressed triumphalism fuelled by an impressive number of professional and academic converts – among them Anglican Archdeacon of Chichester Henry Manning (later to become Cardinal Archbishop of Westminster), and Oxford men like Richard Sibthorp; and scions of land-owning gentry, Ambrose Phillipps and George Spenser, and of course, Newman himself.

These new converts to Catholicism joined by artists, writers and professional men (Wardell would soon be one of them) formed a kind of loose brotherhood, knowing and supporting one another, and using their talents and influence to counteract the widespread no-Popery clamour, the insults, and the occasional violence shown towards them. Even Queen Victoria as titular head of the Church of England, was moved to express her regret at the unchristian and intolerant spirit exhibited by many of her people. 'I cannot bear to hear the violent abuse of the Catholic religion which is so painful and cruel towards the many good and innocent Roman Catholics.'[12]

Violent abuse and the intolerant spirit of the general public did not deter William Wardell but it is significant that he waited until he had legally come of age to take his dramatic step. His conversion was most likely resisted by his father, a life-long and much respected member of the Anglican parish of All Saints. So fierce was the sectarian divide in those years that Catholic priests were often banned from entering workhouses to minister even to the Catholic inmates; Poplar may have been one of these. The little evidence we have points to a complete severance of the relationship between William and his parents. Thomas Wardell was not present at his son's wedding, or certainly did not sign the register at the ceremony when it took place three years later. We know the Wardells were a staunchly Protestant family, prominent members of their parish and employees of the Union. In keeping with the anti-Catholic atmosphere of the times, a family rupture would have been inevitable.

William evidently looked upon his conversion as having found what he had sought at the end of a long journey. He adopted the Latin Motto, 'Inveni quod quaesivi'- I have found what I have sought. He later incorporated the words into his family coat of arms.[13]

CHAPTER 4

THE PUGIN CONNECTION

Passing reference to Pugin has been made in previous chapters. As his strong presence continues rather than diminishes throughout the story of Wardell's life, it is timely that we examine the connection between the two architects, and the influence that Pugin had on Wardell's work.

Wardell is sometimes described as having worked in the shadow of Pugin; it is said that his work is 'Puginian'.[1] A leading authority in Australia, Ursula De Jong, argues that it is more accurate to say that the 'practices of Wardell and Pugin were complementary.'[2] Certainly Wardell learnt much from Pugin but he put his own stamp on his Gothic Revival churches, adapting the style in accordance with local conditions, climate, and financial restraints. He was, in Vincent Wardell's words, 'his own man'.[3] Only on cursory inspection would a visitor be mistaken in thinking that a Wardell church had been designed by Pugin, or that a Pugin church was by Wardell. The character of Wardell's churches suggests solidity and sound construction, Pugin's tend to be slender and more ornate. This was not always Pugin's own judgement of Wardell's work. He thought, for example, that St Mary's Church, Clapham, was 'really quite ludicrous' and a parody of his own work.[4] This strikes us as grossly unfair but a remark in character with Pugin's directness. He was well known for such impetuous judgements and he could - and often did - alter his opinions within a short time. As Hill notes, 'his moods could change like scudding clouds.' - and again, 'his criticisms were never spiteful and his praise was quick and generous.'[5]

Augustus Welby Northmore de Pugin (1812-1852) to give him his full name was the best known, the most articulate and the most influential advocate of the Gothic Revival. Sometimes described as the father of the Gothic Revival, his churches in that style can be found all over England and parts of Ireland. There are several examples of his work in Australia.[6]

His influence on nineteenth and early twentieth century ecclesiastical building was extraordinarily pervasive. Leading architectural historian, Nicholas Pevsner, has pointed out, '[Pugin] saved much English church architecture of the later nineteenth century from the slackness of discipline which one meets so often in secular buildings.'[7] Sir Kenneth Clark echoed this opinion in his celebrated study of the Gothic Revival, 'We have only to think of the average church before Pugin's time to see how far his work is removed from the drab and flimsy buildings of his predecessors.'[8]

The prolific extent of Pugin's creativity – his hundreds of architectural plans and buildings, his designs for liturgical vestments and sacred vessels, his published written works, his lecturing, and his encyclopaedic knowledge of mediaeval artefacts meticulously recorded - remain an authoritative resource even today. Surprisingly, this extraordinary output was accomplished within a sadly short life. If judged in volume alone his creativity would compare favourably with the output of other more celebrated short lives – composers, Schubert and Mozart; writers, the Brontes and Jane Austen; poets, Keats and Shelley. All of these, like Pugin, were dead either before or close to the age of forty. Pugin's energy, his frenetic travelling, and his Spartan regimen injured his health and led, in his last year, to confinement in a mental institution.

His biographers recount how Pugin, at the height of his creativity, explained to a puzzled client why he did not engage a clerk to help him; Pugin was said to have replied, 'A clerk? I would kill him in a week.'[9] His only pupil, John Hardman Powell, who later became his son-in-law and business partner, is reported to have lasted longer than a week, nevertheless he gave way under the strain and had to retire for prolonged rest. In view of this experience it may be accounted fortunate for us that Wardell was not - as some have believed - a pupil of Pugin. When Pugin died in 1852 his doctor famously remarked that although only forty, 'he had done a hundred years' work.'[10] And much of this work was anonymous.

If all this sounds disconcerting, beneath the dogmatist and the rough exterior, Pugin had a kindly nature. He was a loyal but demanding friend; he was a witty and inspiring mentor; and although his acts of charity towards those less fortunate than himself remain mostly hidden, those that we know about suggest that many others may have gone unrecorded.

He eschewed pomposity and had no patience with snobbery, hypocrisy, or humbug. His many friends admired him, his many friendships were deep and lasting. As his latest biographer notes, 'In the thousands of letters and papers that survive to document his life in his own words, and those of others, there is no suggestion, for all his impetuosity and his sometimes exaggerated emotionalism, that he ever did a mean or cowardly thing.'[11]

From all this we can readily understand how Wardell, an up-and-coming architect, a recent convert, and an enthusiastic upholder of the Gothic Revival for ecclesiastical building, could have fallen under the spell of Pugin and his companions. Pugin, in turn, was said to have been impressed by Wardell's study of Gothic architecture and his enthusiasm for the Revival.

We do not know precisely how and when Wardell first met Pugin, but probably they were introduced sometime in 1843 or 1844, around the time that Wardell became a Catholic. Pugin was by then a public figure and an eccentric celebrity; Hill describes him as 'a household name'[12] known as much for his controversies as his churches. His letters and exploits were seldom absent from the newspapers. So well known was he that Robert Browning made reference to him in his long poem, *Bishop Brougham's Apology*:

It's different preaching in Basilicas,
And doing duty in some masterpiece
Like this of brother Pugin's, bless his heart!
I doubt that they're half baked, those chalk rosettes,
Ciphers and stucco-twiddlings everywhere;
 It's just like breathing in a lime-kiln: eh?

Even allowing for Browning's obscure references in his poetry the joke here is clear to us: Pugin would not have been guilty of half-baked, chalk rosettes.

On 1st September 1846 Pugin's crowning work, St Giles in Cheadle, Staffordshire, was ceremoniously opened and dedicated to the accompaniment of much pomp and widespread publicity. The glittering guest list on that occasion included leading architects George Gilbert Scott and Charles Barry; artist friends Clarkson Stanfield, the Scottish painter John Roberts and John Rogers Herbert; from France the Comte de Montalembert and Adolphe-Napoléon Didron, editor of *Annales Archéologiques*; and clergy including Wiseman, Newman, and Archbishop Polding from Australia. At that time Wardell, aged twenty-three, had already designed churches on which Pugin had contributed, and was recognised as a leading Gothic Revival architect. Such an occasion would have been a glittering opportunity for him; many of the guests were to become his close friends. There is good reason to believe therefore that he too, would have been invited to attend the festivities, We know that Wardell kept among his papers a three-page illustrated article on St Giles which originally appeared in the *Illustrated London News* in January 1847. This may not prove he was there but it would seem a natural memento for someone who was. His attendance would also explain how he came to meet Newman (if he did) and Archbishop Polding, for the first time. Also present on that occasion was Robert Willson, a Nottingham priest, brother of an architect and antiquary, who would shortly be appointed the new Bishop of Hobart. Bishop Robert Willson, like his brother a confirmed Gothicist, would, at a much later date, invite Wardell to design the cathedral for his Tasmanian diocese.

The grand opening and blessing of Pugin's church at Cheadle was clearly a significant professional occasion for the young, ambitious Wardell and it is difficult to believe that he would not have been present, surrounded by leading figures in the Gothic Revival. Many commentators regard St Giles as the most beautiful Catholic parochial church in England. Newman, who attended the opening but did not stay to attend the banquet given by Lord Shrewsbury at nearby Alton Castle, thought St Giles was 'the most splendid building I ever saw' and the Blessed Sacrament chapel 'Porta Coeli' – the gate of Heaven. This is in direct contrast with his general opinion of Pugin, 'He does not talk or write like a sensible man ... is narrow minded and a bigot. He identifies love of Gothic art with orthodoxy, and love of classical or ancient Italian art with heresy. He has also said that the sight of Rome is a trial...'[13]

Pugin was the Earl of Shrewsbury's principal architect and the most prominent member of his circle of architects, artists and craftsmen, who found themselves struggling to make good their careers which suffered after their conversion to Catholicism. Wardell was surely among those who journeyed to nearby Alton Towers, the Earl's castle largely restored by Pugin. An additional, intriguing link with Shrewsbury is the fact that Robert Butler, Wardell's future brother-in-law was the Earl's Factor on the estate; it is tempting to imagine, therefore, that he introduced William to his sister, Lucy, on that occasion – they were married a year later.[14] There is little doubt that Wardell, when he became a Catholic, also benefited from the Earl's patronage.

Although Wardell, as we have seen, was not a formal pupil of Pugin's he was a pupil in a looser sense, and like all conscientious disciples he absorbed what he needed from the master and then added his own character and ideas to what he had learned. He was, in his grandson's words already quoted, 'his own man'.

Both Pugin and Wardell believed that the purpose of their ecclesiastical work was to increase and reflect the Glory of God on earth through beauty and fidelity to the principles of Gothic style. They both practised architecture as a religious vocation. Both men upheld the ideal of the architect, 'to do their utmost for the service of God's holy religion.'[15]

Drawing Pugin and Wardell together and providing an aegis for other artists and craftsmen and influential Catholics, was the saintly figure of John Talbot, 16th Earl of Shrewsbury. Largely because of his modesty, his friendliness, his piety and philanthropy, he was popularly known by various epithets, 'Good Earl John', 'The saintly Earl', or 'The millionaire saint.'[16] As England's leading Catholic layman he became the first Catholic peer to take his seat in the House of Lords since the Reformation. Of the many Catholic churches designed by Pugin and paid for in whole or in part by Shrewsbury, the most significant include: St Barnabas Nottingham, St Mary's Cathedral Killarney; St George's Cathedral Southwark (bombed in WWII and then re-built); St Chad's Birmingham; St Giles Cheadle and St Augustine's Ramsgate. The extent of Lord Shrewsbury's munificence is not known, but his total expenditure to charity and ecclesiastical buildings was calculated to have exceeded £500,000: a sum in today's values that may be multiplied by at least ten.[17]

Shrewsbury's patronage and church building should be seen in relation to the discrimination that Catholic professional men experienced at that time. Although the last of the Catholic Emancipation Acts had been passed in Parliament in 1829 and Catholics were entitled to be treated as equal citizens, to receive the same education, and to have the professions opened to them for the first time since Henry V111, there remained a deal of anti-Catholicism fuelled by opponents of the Act. The cry of 'No Popery' was again heard throughout the land. Catholic churches had their windows broken and the newly installed Archbishop of Westminster, Nicholas Cardinal Wiseman, and the Pope, were burnt in effigy. Adding further fuel to the fires of Guy Fawkes Day and beyond, was the inflammable tinder of the Oxford Movement and the notorious Tracts which led to Newman's reception into the Catholic Church in 1845. In this volatile atmosphere recent converts like Pugin and Wardell found themselves excluded from major contracts. How fortuitous, then, was the patronage of the 'saintly earl'. His willingness to contribute towards the costs of so many new churches following the re-establishment of the Catholic hierarchy and a rejuvenated Church, led to an unexpected cornucopia of commissions from which Pugin and, we suspect Wardell, greatly benefited.

There are odd similarities between Wardell and Pugin, both in their lives and works; one cannot write about the former without also studying the influence of the latter. The two lives converge at certain points like images glanced through the right and left-hand lenses of a pair of binoculars. The metaphor, however, should not be overstated; the differences between the two lives are there too. In Newman's judgement Pugin was 'unable to understand or admit anything but his own narrow view of things.'[18] In contrast, although Wardell maintained strong architectural principles, he was courteous and more conciliatory. He was not a rigid neo-Gothicist as Pugin was; he embraced other styles for many of his domestic and public buildings. When we walk into one of Pugin's major churches we are conscious of confronting a work of art; an exhibition of Gothic design. Even Newman as we have already noted, was impressed with the splendour of decoration in St Giles, 'coloured inside in the most sumptuous way.'[19] Wardell's Gothicism, by contrast, while conforming to the same principles, was less confronting and capable of compromise.

Pugin worked himself to death at aged 40; Wardell although working unstintingly throughout his life so that he had periods of illness, had attained almost twice the span of Pugin when he died.

These differences aside, the similarities in their lives are worth remarking upon. Pugin was a very public, crusading Catholic, and was variously accused of becoming a Catholic merely because of his love for Catholic art and architecture. To which accusation he answered, 'I learned the truths of the Catholic religion in the crypts of the old cathedrals of Europe. I sought for these truths in the modern Church of England, and found that since her separation from the centre of Catholic unity she had little truth, and no life; so, without being acquainted with a single priest, through God's mercy I resolved to enter His Church.'[20] This might also be said of Wardell; although as resolute as Pugin, he strikes us as more introvert and self-effacing. They were both pious and prayerful throughout their Catholic lives and both incorporated chapels in their homes where Mass was said whenever possible and where they retired to pray daily. Wardell, as we have seen, flirted with the prospect of a maritime career; and Pugin adopted the guise of a yachtsman and sailed his own small boat in the Thames Estuary and North Sea, to the point of being wrecked on one occasion.

But it was in their advocacy of the Gothic style for ecclesiastical building, that the two architects were most closely united. Wardell would have been familiar with Pugin's influential writings, notably *Contrasts* in which the author compares 'the noble edifices of the thirteenth and fourteenth centuries with similar buildings of the present day; showing the present decay of taste'; and his *True Principles of Christian Architecture* which was published in 1841 when Wardell was eighteen and studying architecture in Mr William East's London practice.

Pugin had been appointed Professor of Ecclesiastical Antiquities at St Marie's College, Oscott, and *True Principles* is a compendium of his inaugural lectures. There, he sets out his two great rules of design, *1st*: that there should be no features about a building which are not necessary for convenience, construction, or propriety; *2nd*: that all ornament should consist of enrichment of the essential construction of the building.' And then he adds: 'The neglect of these two rules is the cause of all the bad architecture of the present time. Architectural features are continually tacked onto buildings with which they have no connection, merely for the sake of what is termed effect.'[21]

Wardell would have read these published lectures and would have absorbed their teaching: in this sense, and this sense only, he could be said to have been a pupil of Pugin. He believed, as Pugin preached, that 'A pointed Church [i.e. Gothic] is a masterpiece of masonry. It is essentially a stone building; its pillars, its arches, its vaults, its intricate intersections, its ramified tracery, are all peculiar to stone, and could not be consistently executed in any other material.'[22] Forty years later Wardell shows, in a letter to the Anglican Archbishop of Perth, that he had not abandoned these Pugin principles. Concerning his plans for St George's Cathedral he argues in a letter to the Anglican Archbishop of Perth that although he agreed reluctantly to build in brick as requested, 'it would have been infinitely better to have been built in stone.'[23]

Both architects were consistent in holding that Gothic was the highest development of architecture and the true Christian style.

So, were Pugin and Wardell guilty of being too emotionally attached to the Gothic style and furnishings? Were they mere aesthetes, 'Romantic Catholics' placing the externals above the essential truths of their religion? Pugin refutes these charges that were made at the time and we cannot doubt that Wardell would have wholly agreed:

> 'The Mass, whether offered up in a garret, or a cathedral is essentially the same sacrifice; yet who will not allow that, when surrounded by all the holy splendour of Catholic worship, those august mysteries appear ten times more overpowering and majestic? May we not confidently hope that, whilst the senses are wrapped in ecstasy by the outward beauty of holiness, the divine truths will penetrate the soul thus prepared for their reception.'[24]

These convergences constitute the Pugin-Wardell Connection. To ignore or minimise them would result in only half the story of Wardell's life and work being told. Wardell's work had been overshadowed by Pugin, but when Pugin's productivity diminished, leading to his incarceration and death in 1852, William Wardell came to prominence, the preferred Catholic architect and in a sense Pugin's heir and rival. But Pugin would remain dominant in the history books, while Wardell would be largely forgotten in England.

It is when we come to consider the personal relationship between Wardell and Pugin that the extent of their much-quoted 'friendship' is open to question. Pugin's latest and most scholarly biographer, Rosemary Hill, doubts that they were much more than professional associates, and occasionally they were even rivals. Hill argues that the relationship was 'not close, or indeed warm.'[25] As evidence she states that in all the thousands of extant letters which Pugin wrote to Hardman, and which the biographer has studied, there is no reference to Wardell to be found. Hill suspects that descendents, proud of Wardell's reputation and achievements, would have had a natural desire to exaggerate the connection with Pugin the more to enhance the reputation of their famous forebear.[26] While Hill's scholarship is unassailable and therefore her views require our respect, there is, however, a case to be made to the contrary - circumstantial as it might be.

Firstly, Wardell writes of Pugin's 'eloquent and elegant advocacy' of Gothic architecture, and describes him as 'our own great master'[27] - not sufficient proof of friendship but surely significant. Secondly the converts to Catholicism in those heady 'Second Spring' years – artists and artisans, members of professions, minor aristocracy and the clergy – formed a close-knit society, known to one another, each giving moral support in various ways to one another. In the anti-Catholic climate of the times they were on the margins of society and in unity was their solace. Wardell was clearly a member of this 'Brotherhood'. Thirdly, we know from extant correspondence that he was on intimate terms with members of the group who were also friends of Pugin – the artists Clarkson Stanfield, David Roberts and John Rogers Herbert; Daniel Rock, the Earl of Shrewsbury's chaplain; James Hope-Scott, lawyer and member of parliament; the famous encaustic tile designer Herbert Minton, and many others headed by Shrewsbury himself, all acknowledging Newman as their spiritual inspiration.

To this writer it would seem strange if all these were Wardell's intimates, yet Pugin who contributed decoration and artefacts to some of Wardell's churches, was not. But whether the friendship was as close as maintained by Wardell's family and accepted unquestioningly by later admirers, is not so clear. Lending support to Rosemary Hill's view we learn that relationships within the group were not always as smooth as we might suppose. Pugin had a bitter falling out with Dr Rock and accused him of 'spitefulness'. When Pugin was suffering his last severe illness,

Rock wrote to Shrewsbury 'recommending William Wardell to finish the work in hand at Alton'[28]: a suggestion which the Earl, remaining faithful to Pugin, rejected.

Unless and until some long-hidden documentary evidence is discovered – and that seems unlikely - we cannot establish whether a friendship existed between the two architects; we can only be sure that they knew one another and that Pugin loomed larger in Wardell's life and experience than Wardell loomed in Pugin's.

Pugin's dominance outlived him and became part of the great Pugin architectural legacy. Wardell was only one among many architects of the 19th century whose work was circumscribed by Pugin's principles. George Gilbert Scott, the most representative figure in the whole Revival wrote of Pugin:

> 'I was awakened from my slumbers by the thunder of Pugin's writings. I well remember the enthusiasm to which one of them excited me, one night when travelling by railway, in the first years of their existence. I was from that moment a new man. What for fifteen years had been a labour of love only, now became the one business, the one aim, the one overmastering object of my life. I cared for nothing as regarded my art but the revival of Gothic architecture.'[29]

As in Scott's experience so we can imagine that Wardell would also have been awakened from slumber by Pugin and inspired by him; it would have been impossible to have ignored him. In Paul Johnson's estimation Pugin was 'one of the most continuously, persistently, and intensely creative artists of all time.'[30]

CHAPTER 5

ARCHITECT IN THE CITY

The young William, barely 20, established his architectural practice in the year 1843 in the name of Wardell and Littlewood[1] and almost immediately produced designs for what was likely to have been his first building success – the Richmond (Surrey) Mechanics Institute. The foundation stone was laid with great ceremony on 26 August, the anniversary of Prince Albert's birthday. The land had been given by Queen Victoria which may have inspired Wardell's choice of the Italianate style, complete with columns and bow frontage. Designed to house a theatre, meeting rooms, museum and library the various Mechanics Institutes were the fore-runners of public libraries 'for the self-improvement of the working man.' The following year the Royal Academy exhibited a drawing of the Institute – considered a mark of commendation, and a significant step in the career of a young architect trying to make his way at such a young age.[2] Notwithstanding this success he had to wait nearly two years for his first church commission, St Edmund's, on the Isle of Dogs (1846). No sign of it remains today in the recent Thames-side development.[3]

Wardell's next church commission came later that same year, and would prove a much more substantial building than St Edmunds, one that would reveal William Wardell as a sensitive artist and a major exponent of Gothic Revival church architecture.

But before we consider 'Our Ladye Star of the Sea', Greenwich in some detail, we alight on the next significant event of a more personal nature in Wardell's life, his wedding which took place on 5 October 1847. William's bride was Lucy Anne Butler, eldest daughter of William Henry and Elizabeth Butler of Oxford.

Lucy's father, was born in 1790, the ninth of ten children. On 13 February 1817 he married Elizabeth Briggs the youngest daughter of a Northampton Alderman. Butler flourished as a prominent, fashionable wine merchant in Oxford with premises in the centre of the town at 2, St Aldate's. Given the address, and his status in the town, we can be sure that he supplied Oxford High Tables with the best Port and Sherry.

One of Butler's two sons, Edwin, ran the wine business in town which allowed William Henry Butler to travel and be an active member of the local Council, becoming Senior Bailiff in 1824, and serve as first Mayor of Oxford for a brief period in 1836 when the new Municipal Reform Act came into effect. The second son, Robert, was appointed Factor at Alton Towers, the estate of the Earl of Shrewsbury. If William was known at Alton Towers by that date, Robert may have provided the introduction to his sister. Lucy Anne, the elder of the Butlers' two daughters, was born in 1821, therefore two years older than her husband.

Another link may have been John Wardell, William's cousin, who was employed as a clerk in the East India docks. He would have had business dealings with William Butler who styled himself an 'Importer of Wines and Spirits'. In this capacity he would have needed to visit the wine vaults close to the docks in Poplar. John's wife was Caroline who, on the 1851 census form, is shown a visitor to the Butler household; from this we can assume she had become a friend of the Butler daughters.

William and Lucy's marriage was solemnised 'according to the rites and ceremonies of the Catholic Church' by the Reverend John Walsh at St Mary's Chapel in the City. Lucy and her brother Robert had become Catholics independently of her suitor, William. Evidently the Butlers were liberal enough not to object to the union on religious grounds because William Butler was one of three names on the Register. The Groom's father, as we have noted, was absent. Some significance may be attached to the appearance of a formal announcement of the marriage which appeared in *The Times* two days later. 'On 5th inst., at St Mary's Catholic Church, Moorfields, by the Rev. John Walsh, W. W. Wardell Esq., architect, to Lucy Anne, eldest daughter of W. H. Butler, Esq., of Headington, Oxford.' There seems to be, in that terse public statement, something proud and challenging, as if Wardell wanted the fact to be known publicly, defying opposition. Wardell's inclusion of his profession in the marriage notice suggests also that he was promoting his recently acquired status, and his address is given as 27 Bishopsgate Street. Evidently William was judged by the bride's parents to be a suitable match for Lucy, being a member of a respected profession, and one who had prospects of advancement. However, as a Catholic he would face the kind of prejudice that beset many other professional men who became Catholics at that time, as instanced by Vincent Wardell's statement that his grandfather's conversion 'involved for him the cessation of many old friendships [and] the loss of a rich inheritance from a close friend of his father.'[4]

Fortunately for William Wardell, although he may have lost an inheritance and some friends, and business opportunities, he would find others, and would be welcomed into an influential circle of fellow converts and clergy. Newman's so-called 'Second Spring' would result in an unprecedented requirement for new ecclesiastical buildings the length and breadth of England. Architect William Wardell, like Pugin, was therefore ideally placed to benefit from the new opportunities on offer.

Those latter years of the 1840s were extraordinarily creative and busy ones for Wardell. His reputation grew as a leading ecclesiastical architect in the Gothic tradition, second only in popularity after Pugin. England's most exacting critic, Nicholas Pevsner, concedes 'The most note worthy church work of the 40s is by Pugin ... and his immediate follower, Wardell...'[5]

The exact number of Wardell's buildings - churches, presbyteries, convents and schools - designed in the twelve years between 1836 and 1858 (when he left England for Australia) is difficult to calculate accurately. One authority, Ursula De Jong, suggests 'over thirty' and we may take this as a reasonably accurate figure.[6] Certainly there are fourteen of special interest which still exist and are in use today; and we can account for at least eight others that have been destroyed by enemy action in World War II – among them, St John the Baptist, Hackney (1847), St Mary and Joseph, Poplar (1856) and St Edmund, Isle of Dogs (1846). There are others which have been demolished by those who chose to replace them by larger and more modern constructions: St Richards, Chichester, (1856) and Our Lady Help of Christians, Kentish Town (1849), Our Lady and St Joseph, Kingsland, and St Boniface's in London's East End.[7] Even the best examples still in use have been altered or added to in the years following changes to the Catholic liturgy after the Vatican II Council. As Wardell's grandson, Vincent Andrew Wardell, ruefully explains in his memoir, architects are wise enough not to sign their work publicly because they know that others will add to or change their work in future years and thus the reputation of the original architect will invariably suffer.[8]

From the list above it follows that Wardell's church commissions filled his order book around the time of his courtship and marriage – although no particular significance need be read into that - it is unlikely that the Butlers were able to give him work. For the bestowing of influence we must look elsewhere. Because there were often delays in construction, and the design of one church often overlapped with another, it is difficult to consider his churches in strictly chronological order, thus they are grouped according to their relationship, one with another.

One of Wardell's first commissions was the little church of St Birinus in Dorchester-on-Thames opened in 1849. Visitors may be forgiven for misjudging its age for it gives the appearance of being much older than its one hundred and fifty years. Set as it is in a rural churchyard overshadowed by trees, with a simple Gothic exterior of ragstone, a steep-pitched roof of weather-worn slates, and a substantial stone porch, it invites us to enter. The interior exudes an atmosphere of rural peace, and surely it has one of the most aesthetically pleasing, romantic interiors of any Wardell church. The carved Rood Screen and painted chancel roof and various Gothic furnishings show the strong influence of Pugin. This is a church that draws us in and will not easily let us depart.

While in the course of construction Wardell found the church and its architect the subject of critical correspondence in *The Tablet*. The writer, signing himself simply as J.R., begins by hoping that 'Mr Wardell will doubtless receive in good part the remarks of an admirer of his noble art.' He then goes on to complain, in a lengthy piece pompously written, that the 'windows of the nave are somewhat deficient of height' and that 'the chancel has not its eastern aspect'. He also complains of the stone-work 'unfortunately constructed in the spirit of economy', and the 'finished capital of the right hand column of the chancel arch is of a truly mean character'. His strongest disapproval is reserved for 'the prominence of the most unsightly chimney attached to the sacristy... the most un-ecclesiastical, ugly and lumpy abortion that could have been presented to the eye.'[9] There is much more in the same spirit and one suspects from the writer's use of technical terms that it was written by a disappointed rival architect.

The following week Wardell's spirited reply is printed together with two other letters of support for him. He answers each point of criticism in turn, complaining of the injustice of criticising a building before it is half finished: 'J.R. might as well gain admittance to a painter's studio, and see the first "laying in" of a picture, and then comment on the aerial perspective.'[10] Addressing the criticism of the sacristy chimney Wardell concedes that J.R. is entitled to his opinions, quoting: 'De gustibus non est disputandum; but it is most certainly ecclesiastical, and I am borne out in this by the practice of one of the greatest Christian architects, Mr Pugin'.[11] Wardell argues that it is preferable to have a chimney to carry away the smoke, rather than use a charcoal or coke burner that would asphyxiate the worshippers or damage the gilding and other decorations.

> 'With regard to the aspect of the chancel, he is quite mistaken; it is as nearly as circumstances will allow and I don't think it is three points of the compass out. There are numerous ancient instances of the orientation varying from north-east to south-east'.[12]

Wardell's letter concludes by asserting that 'J.R. should consider twice, before he writes in a manner, which to say the least, is likely to ruin the professional reputation of a young man.'[13]

The letter is the first of several published letters that Wardell would write in answer to criticism throughout his long career – always polite, measured, but incisive in argument. In addition it shows Wardell's admiration for Pugin as a mentor, and we note his use of the nautical term 'three points of the compass' which would be unfamiliar to anyone who had not been trained in navigation.

Of the two letters of support and in praise of the church one, who admits he knows Wardell speaks of him as 'a good and amiable architect' and reports that many believe of St Birinus that when it is finished 'nothing of the kind in England will surpass it in beauty'. The second letter-writer states he does not know Mr Wardell so cannot be biased, but he believes J.R.'s criticism was 'most unjust' and that St Birinus 'is already a perfect model and a gem of Christian art.'[14]

St Birinus is clearly an early work – the plans are dated 1846 - and although certainly not pastiche, it is eclectic taking its inspiration from

Wardell's close observation of pre-Reformation village churches. The exterior already shows signs of his individuality. The west gable, the porch, and little bellcote are recognisable characteristics of Wardell's smaller churches. The tracery is curvilinear of the 14th century and happily the carved and decorated Screen survives in place, as does the Sanctuary and Altar, seemingly unaffected by present day liturgical changes. The custodians of St Birinus must know they have a rare jewel of a parish church which, like the village itself, has to be preserved and cherished as the architect and the original patron, wealthy Catholic yeoman John Overy, intended. Wardell's faithful historicism conveys a sense of continuity with the medieval abbey close by – the more popular focus of interest for tourists. But St Birinus, a poorer cousin to the abbey maybe, nevertheless by its intimacy and its modesty enriches and sanctifies the neighbourhood.

Other Wardell churches designed in the second half of the 1840s include the three already mentioned that were damaged and then demolished during the War.

Those last years of the decade saw the maturation of Wardell's talents. Three major London churches designed in close succession enhanced the architect's reputation and led to further commissions. Happily the three - Our Lady of Victories, Clapham Common, in south London; the church of the Most Holy Trinity, Brook Green, Hammersmith, and Our Ladye Star of the Sea, Greenwich,[15] are still in use today and well attended. The foundation stone of the Clapham church was laid in August 1849 and opened with great pomp by Cardinal Wiseman on 14 May 1851. Most likely the architect was among the invited guests by right, but he also became a personal friend of the French Redemptorist Rector, Father De Hels. In spite of a stone being thrown at the Cardinal's carriage, the opening ceremonies were not marred by anti-Catholic demonstrators in the vicinity.[16]

Like the other two churches under consideration, Clapham was built of Kentish ragstone with Caen stone dressings. The nave has six bays and north and south aisles, the main entrance being under the tower on the north side. The height of the tower and spire rises to one hundred and seventy-two feet, wherein hangs a complete set of bells, popularly claimed to be the first peal sounded in a Catholic church after the prohibitions had been lifted.[17] The church was much altered and added to in later years. J. F. Bentley, a parishioner most famous for his Westminster Cathedral, was responsible for adding the transepts, the richly ornate chapel of Our Lady of Perpetual Succour, and the adjoining clergy house. But the sanctuary and its furnishings, and the nave and chancel – the former with its fine timber vault and the latter having a gilded rib stone vault - are Wardell's work and can be admired for their spaciousness, light, and elegant proportions. Pevsner, that exacting critic, described St Mary's as 'one of the best Victorian churches in South London.'[18] Pugin in a pique was less enthusiastic as we have already noted.[19]

Seven days before the grand opening of Clapham, Cardinal Wiseman was at Brook Green in Hammersmith on 8 May to bless and lay the foundation stone of Wardell's second church in this trio. The Church of The Holy Trinity forms part of a pleasing complex of buildings, including a presbytery (rebuilt in 1963) and the St Joseph Almshouses also by Wardell. Our view of the almshouses, fronted by a private lawn giving them the appearance of a peaceful medieval college, is now blocked by modern flats fronting the street. As at Clapham, Wardell chose Kentish ragstone with Caen stone dressings for the exterior but in contrast to Clapham the tower is placed in the south-west corner. The spire, included in Wardell's design, was not erected until after he had left for Australia and was modified by J. A. Hansom in 1871. The overall style is broadly Decorated Gothic of the 13th century. The nave has six bays and north and south aisles with a second (outer) aisle on the north side. The fine timber roof is supported on brackets resting on corbels enhanced by carved angels playing musical instruments alternating with kings bearing gifts. Above are the clerestory windows. The chancel is two bays deep and paved with Minton's encaustic tiles. The splendid east window depicting the Passion of Our Lord is a Pugin design, the stained glass by Hardman, Pugin's business associate.

The Brook Green Church has been described by one historian as 'the high water-mark among Wardell's London churches'[20] and certainly it seems the most satisfying in terms of unity of design and proportion. For any visitor intending to make a Wardell pilgrimage in England, the Church of The Holy Trinity, Brook Green, should be at the top of the list.

The third church in the trio, 'Our Ladye Star of the Sea', Greenwich, is both a joy and a disappointment in equal measure – disappointment because of the clumsy alterations in the 1960s which, it must be said, the present incumbent hopes to rectify when the high costs of restoration can be met.[21]

The best approach to the church is up a winding residential road, Crooms Hill, which leads out of the town centre with Greenwich Park on the left. The handsome exterior of Kentish ragstone gleams in the sun and is dominated by the great corner-buttressed tower and spire rising proudly above the surrounding rooftops. It is said that it was once a landmark for ships on the Thames close by. *The Tablet* reported that 'many an aged eye was dimmed with joyful tears' when the cross was raised to the top of the spire in 1849.[22] The main entrance of the church is at the base of the tower. The roof suffered damage during the war and on a second occasion during the Great Storm during the night of 16-17 October, 1987 when a neighbouring chimney stack was dislodged and crashed through the slates.

It is only when we enter the church that the first disappointment becomes apparent. A false plywood ceiling has been added in the nave, hiding the fine decorative wooden vault similar in style to Brook Green. Somewhere above the plywood are carved, heads depicting old Kentish saints and bishops, each protruding from their placement on the end of the corbels. The loss of this treasure, an integral part of the design, is calamitous, excused in the destructive 1960s on the grounds of insulating the church against freezing winter temperatures. For a similar reason cork tiles were laid on the original floor tiles throughout the church. Stencil work on the aisle walls from designs by Pugin has been painted over. At time of writing the rendering on the pillars has been removed and happily the Purbeck marble of the six bays has been restored.[23]

Having listed the dismal changes, which were inflicted by an over-zealous committee in 1965 wishing to mark the Golden Jubilee of their semi-retired parish priest, there is still much left to admire. For example, the elegant proportions of Wardell's design, the beautifully-formed arches and sense of light and space looking down the nave towards the chancel. Unusual in a large modern Catholic church is the Rood Screen carved in polished Caen stone, and surmounted by a painted Holy Rood - a welcome survival which even the most ardent iconoclast would have found difficulty in destroying. The carved stone altar rails and the High Altar itself, both designed by Wardell, happily survive in spite of the liturgical changes. Similar carved stone is used for the baptistery enclosed in the north aisle at the west end of the church. The superb stone statue, Our Ladye Star of the Sea, based on a sketch by Pugin was carved by his chosen craftsman, George Myers. Although all the internal decoration, stencilling, altar and tabernacle in the Blessed Sacrament Chapel, and the rich decoration in the chancel is commonly attributed to Pugin himself, but this, according to one authoritative opinion, is wrong and that much of the detail was, in fact, carried out by his son Edward.[24] By 1851, when the church was completed after long delays, Pugin was prematurely worn out; he complained of insomnia, nightmares, constipation, and depression, and his hitherto frenetic activity was often interrupted by days of sickness.[25] The same view holds that he only managed one visit to the Greenwich church – his last to any of his many churches.

The completion and opening of Greenwich, like so many other Wardell churches both in England and Australia, was subject to long delays on account of a shortage of funds. Father Richard North, the first pastor of the original chapel which served the district at the time, found difficulty in attracting money. In his letter of appeal published in *The Tablet* he first compliments his chosen architect, Wardell, and then gently chides him for his exclusive dedication to Gothic. 'The amiable architect of Our Lady's Church has followed out his own partialities in the style of building, and in his very first effort has sprung to the foremost rank among the gifted of his profession.'[26] The writer then argues that 'these medieval forms no more indicate faith and exclusive reverence for Religion than do square-toed shoes imply sound morals...the soul of Religion is not in burnt bricks or quarried stones'.[27] Here we see the crux of the argument aimed at the Gothic Romanticists that was known as 'the battle of the styles'. Fortunately Father North had a sense of humour and, in spite of his reservations, admired his architect - and Pugin too, since he was responsible for so much internal work. His humour is best shown towards the end of his long letter when he urges the Catholics of London to visit the site of the church and we can well imagine the poor priest desperate to find sufficient money to pay for Wardell and Pugin's extravagant ideas. He urges visitors: 'If you approve, do not clap your hands and pass by, as some have done, but put them in your pockets. Ask any Master of Ceremonies: the proper place for the hands on visiting an unfinished church is the pockets.'[28]

Father North found his money eventually, some of it via a grant from the Admiralty, but perhaps a more significant portion coming from a rich merchant family, the Knills. Providentially for Father North, Pugin married his third wife, Jane Knills, on 10 August 1848 – 'A first rate Gothic woman'[29] – and clearly it was through her connections that the money was raised and Pugin was commissioned to complete the interior. But neither should we forget the more humble contributions of those poor passers by who were persuaded to put their hands in their pockets. Now the good priest lies in peace in a tomb shared with his brother, Canon Richard North, at the foot of the High Altar.

Father North's gentle reproof of Wardell's insistence upon the Gothic style for churches was only the tip of more forceful objections being expressed by clergy mainly on the grounds of a perceived excessive cost. Others, like Newman, thought that Classical styles were equally valid for new churches. Both Wardell and Pugin responded with equal force. Pugin led the way with a long letter – one of many in the Catholic press over time – this one in the *Weekly Register*, arguing that the House of God should stand as 'the one green spot in the desert of the modern city'.[30] He quoted as justification for the possible higher costs the gospel story of Mary Magdalene anointing Christ with a precious ointment to the objection of Judas. Wardell's argument which appeared in the same publication but on a later date, refers to Pugin as 'our great master.'[31] He writes that although Pugin's 'eloquent and animated advocacy of Pointed Ecclesiastical Architecture leaves but little to be said on its behalf,' he nevertheless sets in detail how the costs of building in the Gothic style are actually lower than building in the Classical style – then favoured as an alternative.

> 'In a Classic edifice, the walls are required to be at least one third as high again as those of a Gothic building containing the same number of cubic feet; and in this item not only to be taken into account is the extra height, but an increased thickness; for a high wall with a flat roof, where the lateral thrust is enormous, requires to be much stronger than a low one with a high roof where the pressure is almost vertical.'[32]

Wardell continues listing the technical differences and adding tables of comparative costs accompanied by drawings of the two styles to emphasise his points. He even calculates the cost of Gothic when compared to 'Pagan, nineteenth Century' – his term for the Classical revival - as it relates to the cost per worshipper accommodated. He finds that when building in Gothic the cost per head averages £8.10s; whereas the cost when employing 'Pagan structures' is £23.10s. per head. Such exhaustive, calculated detail in argument is a recurring feature in Wardell's approach to his work. He would have been a formidable if rather dogmatic, opponent.

Continuing in his *Weekly Register* article covering over four pages of double columns, he does not limit his reply to comparative costs alone. He turns his attention to a broad defence of Pointed Christian Architecture - and here he is using Pugin's preferred name for Gothic. His arguments are strangely familiar to us today, a period when the 'battle of the styles' has returned, dividing the modernists from the more traditional. The views of Newman, a recent convert and public figure, typified a common antipathy to Gothic then as now: 'Gothic is now like an old dress, which fitted a man twenty years back but must be altered to fit him now ... I wish to wear it, but I wish to alter it.'[33] Wardell, irenic in argument and less belligerent than Pugin, wrote:

> 'A church architect has the Church's honour, and the Church's traditions to consider, before he thinks of gaining a reputation for novelty of invention, which after all is a questionable one; and we should indeed be cautious how we deform the temples of the Most High for the sake of being original and ingenious... Pointed architecture has, on Catholics at least, an indisputable claim, as being the perfection of that system invented by the Church herself as the best adapted to symbolize her faith and teaching, as well as the most practically useful for her ceremonies and worship; our notions and ideas of a religious edifice have thus become bound up and associated with a Gothic church and why should they be violated?'[34]

Wardell concludes his long article with a plea not to condemn the Revival 'which has done so much good in bringing to men's minds the "days of faith;" and which has such probabilities of further benefit,' and ends by reminding readers that we should all labour for Christ's Glory, "...quoniam confirmata est super nos misericordia ejus, et veritas Domini manet in aeternum.'[35]

Wardell's prominent appearance in the widely read *Weekly Register* was welcomed by his friends and fellow Gothic Revivalists. One of these was the influential Dr Daniel Rock, chaplain to the Earl of Shrewsbury, antiquary and scholar of the English Catholic tradition. Rock wrote enthusiastically to Wardell:

'I have but this very moment finished reading your few remarks on Gothic Architecture and so delighted am I with them that I must not let a post go by without thanking you for this pleasure they have given me...I hope you will often give the world the benefit of your thoughts upon the subject.'[36]

Daniel Rock was also, for a time, a close friend of Pugin's, assisting at many openings and blessings of Pugin's churches. He received into the Church Pugin's second wife, Louisa, with great pomp and colourful celebration. Another clue to Wardell's friendship with Rock is an affectionately signed copy of one of Rock's works, *Articles of Doctrine*, which survives in Wardell's library. This also suggests that the young architect was a respected and admired member of that exclusive circle of artists and writers and leading Catholic laymen who supported one another, described in a previous chapter as gathering around the Earl of Shrewsbury espousing the Gothic Revival.

The four Wardell churches described in detail above and designed - if not completed - before the middle of the century, together with those churches which were casualties of the War, represent a remarkable achievement for a young and comparatively inexperienced architect. The ecclesiastical historian, the late Denis Evinson, chose Brook Green as a high water mark in Wardell's London churches.[37] If we take this literally to mean that nothing designed before or since reached its standard, the opinion is questionable. But if the term only means that Holy Trinity Church was a significant high point in Wardell's works to that date, we can more readily agree. It certainly shows a remarkable degree of maturity and unity of style and an understanding of mass and structure - although this may be said also of the Greenwich and Clapham churches. However, it is indisputable that all three of them, plus St Birinus, were built under the shadow of Pugin, and owe much to his influence.

But in the decade from 1850 onwards, Wardell emerges from the Pugin shadow and, in his grandson's words, became 'his own man'.

CHAPTER 6

THE HAMPSTEAD YEARS

The newly-weds, living 'above the shop' at number twenty-seven Bishopsgate Street within the City of London, may have found the address convenient enough for business, but evidently they considered it not salubrious for a young family with rising aspirations. William was ambitious and was fast developing important friendships within the artistic community, many of whom tended to favour the outer suburbs - most notably among these was Hampstead, adjoining the sprawling, as yet untamed Hampstead Heath. It was to Hampstead that William brought his family shortly after the birth of his first child, Mary Lucy, registered in Hackney in 1848.[1]

The Wardells' embossed writing paper gives The Green Hill, Hampstead, as their exiguous postal address a year after their marriage – its brevity indicative of the village nature of the leafy suburb in the middle of the nineteenth century. The name derives from the site of a grass farm on the fringes of the old village, and was originally part of nearby Hampstead Heath. It is known today as 'Greenhill', and refers to a crescent off the south side of the High Street, a short walk south from Hampstead tube station. The Wardells' leased house was in fact a double one in a Georgian terrace of three, built about 1730. It was listed on official documents of the time as 173 High Street, but the number was evidently thought unnecessary on Wardell's printed notepaper. At number 172, next door, lived the eminent marine painter and Royal Academician, Clarkson Stanfield and his large family. He added a studio on his north side of his house which is still there, now converted into a residence. The Wardells' neighbour on the south end of the terrace, occupying a large mansion known as 'The Rookery', was Thomas Jackson, builder and railway contractor.[2] Stanfield's handsome house now named after him, is still there, identified by one of London's singular blue plaques. It stands up high off the road on the corner of High Street and Prince Arthur Road. Sadly, Wardell's house – two houses adjoining - has fallen victim to later development; sacrificed in 1871 to a road named in honour of one of Queen Victoria's children. Prince Arthur Road cuts through what would have been Wardell's garden and house, and divides Stanfield House from the modern block of superior residential units on tree-fronted Greenhill.[3] In the immediate vicinity are more modern houses and apartments built in recent years.

In spite of this recent development Hampstead manages to retain its leafy village atmosphere although a teeming metropolis surrounds it. The residential streets comprise a mix of late Georgian, Victorian, and Edwardian mansions, while some are converted into flats, interspersed with more modern blocks. Hidden from the passing traffic can be discovered a maze of secluded lanes lined with highly desirable, restored eighteenth and nineteenth century houses and cottages favoured by the successful and the famous. These byways have names like Shepherds Walk, Pilgrims Lane, Heath Street and Holly Lane among others, reminding us that Hampstead was a country village one hundred and fifty years ago. When the Wardells lived there, sheep and cattle grazed on the Heath nearby.

Stanfield's son, George, an artist like his father, made two sketches of the Green Hill terrace which he presented to the Wardells on their departure for Australia. They depict an idyllic semi-rural setting. The front view is approached by a laneway and the neat Georgian front doors of Stanfield's and Wardell's houses are side by side. The houses are three-story with a basement, and Stanfield's house shows his studio annex, originally the stables. The studio sometimes doubled as a ballroom where Landseer, Thackeray, Dickens and David Roberts would be among the many eminent guests; Wardell and his family would have been included. Stanfield's old studio, the walls still lined with the original Baltic pine panelling, showed signs of an artist's occupancy as late as the 1960s. George Stanfield's sketch of the back garden shows Wardell's section having an ornate conservatory porch and a leafy garden. On the path, strides a dapper figure, surely meant to depict a fashionable William Wardell himself.

Stanfield first became acquainted with the area when journeying between London and his one-time home in Hertfordshire. He often delayed there to visit his fellow artist, John Constable. In 1846 he rented a summer lodging in Hampstead, and when his fellow Royal Academician, C. R. Cockerell, told him that a house was available he 'snapped it up,' moving to The Green Hill in the Autumn of 1847.[4] This coincided with the year of Wardell's marriage, and just as Cockerell had introduced Stanfield to The Green Hill, it is likely that Stanfield alerted his friend Wardell to the vacancy next door to him. Number 173, the three-storey double house next door, would have been a desirable address for a young family man beginning to establish himself in his profession.

The Stanfield and Wardell families were not only close neighbours, living side by side in the terrace, but very close friends. Stanfield was to describe Wardell as his 'dearest friend'[5] and in his testimonial written when Wardell was about to leave for Australia he stated that 'from our long and intimate friendship few are better qualified than myself to speak of your private worth in your domestic circle'.[6] When Wardell's first son, Edward, was baptised in Hampstead in September 1850, he was given as his second name, Stanfield, in honour of his godfather, Clarkson Stanfield. It is this writer's opinion that Stanfield, thirty years older than Wardell, replaced Wardell's own father in his affections, because of an estrangement following William's conversion to Catholicism. Wardell, long settled in Australia when Stanfield died in 1867 wrote: 'He was a charitable, religious, gentle, truly good man...incapable of pretence or concealment.'[7] Dickens wrote of him, 'He was the soul of frankness, generosity and simplicity, the most affectionate, the most loving and most lovable of men.'[8]

Clarkson Stanfield, born in 1793 - therefore thirty years senior to Wardell - was originally apprenticed in the Royal Navy. He served for a short period under Captain Charles Austen, the much-loved brother of novelist Jane Austen, but his painting talent soon began to assert itself. After his discharge due to a leg injury he found employment as a scene painter at the Royalty Theatre. Those were the theatre days of Macready's and William Barrymore's lavish productions in Drury Lane when the artist responsible for the elaborate scene paintings could be as much a star as the leading actors. Gradually Stanfield gained a 'legendary reputation as a creator of romantic landscape scenery... He was especially famous for his vast moving dioramas'.[9] Although Stanfield maintained a connection with scene painting throughout his life (he painted over five hundred and fifty scenes in more than one hundred and seventy productions in his scenic career), through his close friendship with Dickens he was also famous for illustrating several of his Christmas books - *The Chimes* (1844), *The Cricket on the Hearth* (1845), *The Battle of Life* (1846) and *The Haunted Man* (1847). Dickens dedicated *Little Dorrit* to him on its completion in 1857.

But probably Clarkson Stanfield is best remembered today as one of England's finest marine painters. There are examples of his marine paintings in various British galleries as well as the Royal Collection, the Royal Academy, the V. & A., and the Maritime Museum at Greenwich. In Australia a fine example of his work, a large and dramatic seascape 'Mount St Michael,' hangs in a prominent position in the National Gallery of Victoria. Next to Turner, Ruskin thought him 'the noblest master of cloud forms of all our artists', and explained: 'I cannot point to any central clouds of the moderns, except those of Turner and Stanfield, as really showing much knowledge of, or feeling for, nature..'.[10] Ruskin, an exacting critic, devotes much space to Stanfield in his influential work, *Modern Painters*, but not always without expressing reservations. However, he leaves us in no doubt as to Stanfield's stature as a marine artist: 'Stanfield alone presents us with as much concentrated knowledge of the sea and sky as, diluted, would have lasted anyone of the old masters his life... [he could] carry a mighty wave up to the sky and make its whole body dark against the distant light using nothing more than unexaggerated colour to gain the relief.'[11]

The Stanfields and the Wardells were united not only in their contiguous housing and their love of art, but more importantly by their religion (both being converts to Catholicism). They shared friendships with other artist converts in the Pugin circle. It cannot be stated with certainty that they both knew each other before William moved to The Green Hill but it seems likely and that Stanfield invited him to rent the double house next door to him on a friendly basis as suggested above. A letter from Stanfield dated 13 May 1848, inviting the Wardells to dinner, seems to indicate that they did not live next door to one another until later but it is indicative that the friendship existed earlier: 'Will Thursday instead of Wednesday suit you and Mrs Wardell?...I have asked our good priest Mr Parkinson to meet you but his duties will not let him be with us on Wednesday.'[12] If the Wardells were about to become neighbours and move in next door, or had only just done so, the invitation to meet the priest would make perfect sense.

The Reverend Thomas Parkinson was not officially the parish priest of St Mary's, Hampstead, but assisted the much-loved, aging Abbé Jean-Jacques Morel who was one of over five thousand clergy who had sought refuge from northern France after the Revolution in 1789. The Wardells knew and admired him in his last years – he died in 1852 – and Wardell designed his imposing tomb and monumental effigy which rests in the Lady Chapel in the small south transept. Wardell was also responsible for repairs and restoration work and for the west front bell tower and belfry (Bells in Catholic chapels silenced since the Reformation were allowed to be rung again following the Emancipation Acts of 1829). Wardell chose an Italianate design for his west front to match the interior of the church and the effect is striking, a contrast in white stone with the brick-faced Georgian residences either side in Holly Place. On the right is the presbytery, and on the north side was a convent school and orphanage when the Wardells were parishioners. St Mary's is well worth a visit today, but apart from the monument sculptured in Caen stone and the West Front, the ornate chancel with its barrel vaulting, the imposing baldacchino designed by Adrian Gilbert Scott, the richly painted walls and the Doric pilasters bear no evidence of Wardell's architecture. We can, however, imagine the young Wardell family in their Sunday clothes, perhaps in the company of the Stanfields, walking up the cobbled hill, Holly Lane, as far as Holly Place, and hearing Mass celebrated by the aging and infirm Abbé Morel.

As promising as Wardell's contacts were before coming to live at The Green Hill, undoubtedly it was through Stanfield that his social and professional life was enlarged and enriched even further through meeting Stanfield's important friends, the leading writers, artists and actors of the day who frequently called 'next door'. Those ten years of the Wardells' life in Hampstead must indeed have been rich and happy in friendships and stimulating company. If the few mementos and written records of those years that still exist in the family papers in Australia are any indication, the Hampstead decade proved a treasured memory and a high point in their lives. Through Stanfield they met and became close friends with another convert painter friend of Pugin's, John Rogers Herbert, who painted celebrated portraits of Cardinal Wiseman and Pugin among others, as well as frescoes and other paintings commissioned for the House of Lords. Others who were frequent visitors to The Green Hill included David Roberts, R.A., William Thackeray, Edward Landseer, and the eminent literary critic John Forster, who was Dickens' influential agent and confidant. It is likely, too, that through Stanfield Wardell met, about this time, the rising celebrity John Ruskin, and his first wife Effie. If so the name Wardell would have been strangely familiar to Ruskin. In his autobiographical work, *Praeterita*, he recounts how his father, desirous of John's marriage, arranged meetings with suitable, eligible girls. One of them was a Miss Wardell whose home was in Hampstead - a maddening coincidence and we long to know more. The place or residence and surname surely indicate some connection with William Wardell; but Ruskin makes only a passing reference to her without adding even a Christian name.[13]

Among other close friends were the popular actor-couple Mary Ann and Robert Keeley, as well as members of the Dickens family.

The young Wardell, then aged under thirty with an emerging reputation as an architect, was evidently stimulating company himself, popular and able to match the sagacity of the more celebrated guests. When the Keeleys, popular stars in their profession, were concluding their season in Oxford, Robert, acting on instructions from his wife, wrote to Lucy Ann Wardell in advance of their return to their home in Pelham Crescent, Brompton, inviting the Wardells to dine. In the somewhat extravagant, flowery writing style of the actor, Robert explains how he is staying 'at home' with a cold while his wife has gone to church. He hopes that even though they will not dine on that occasion until 6.o'clock, the Wardells will come early so that 'she will be able to enjoy a good slice of your agreeable companionship.' He then adds, 'Of course the gentleman whose name you bear will have the goodness to understand that all we address to you – the better half – is also inclusive of him, as the major includes the minor.'[14] We do not know who else was invited on that occasion but we can be sure that it was a very entertaining evening.

Of all the illustrious contemporaries associated in the Wardell-Stanfield circle of those years it is perhaps the name of Charles Dickens that attracts our particular attention, for evidently Charles Dickens himself became a family friend.

Dickens had recently bought Tavistock House in the London Square of that name, close to Regents Park and Euston which even in those early days of train travel had been a main-line station for nearly twenty years. The house had been refurbished to the requirements of his large family and included a schoolroom converted into a theatre which he grandiosely named 'the smallest theatre in the world'. It was here that Dickens devoted himself with extraordinary energy to acting and directing. He was so precise and so demanding of everyone around him that it became an obsession, and was certainly an escape from the public's perception and adulation of Dickens - the most celebrated and widely read author of all time, the equivalent in modern terms to a film or pop star. In 1855 he had started writing *Little Dorrit* but then, after only one episode, set it aside to concentrate on his private theatricals – an amateur production of Wilkie Collins' sentimental drama, *The Lighthouse*. In his mammoth biography of Dickens Peter Ackroyd describes how he threw himself into the rehearsals with an extraordinary intensity, and how it 'all sounds rather terrifying for everyone involved'.[15] Dickens's son – also Charles - has left an account of his father's exacting standards at rehearsals. The wind, the rain and the thunder and lightning, so carefully plotted 'had to be done exactly at the word, of course, and on each occasion only, for a rigidly defined time and I could always tell from the very look of my father's shoulders at rehearsal as he sat on the stage with his back to me that he was ready for the smallest mistake and if I didn't wave the flag at exactly the right moment, or if the component parts of the storm were at all backward in attending to their business, there would come that fatal cry of 'stop!' which pulled everything up short, and heralded a wigging for somebody.'[16]

The Wardell family, including Wardell himself, were among those involved on this occasion. Catherine, Dickens's wife, sought the help of Mrs Wardell by sending her a playbill which she asks her to copy 'as soon as possible, our place being very limited indeed and the demands upon it immense'.[17] She then reminds Lucy Ann that 'it will be necessary on the great occasion to arrive before eight as we cannot get anybody into the room after the play begins.'[18] Besides styling himself 'Lessee and Manager, producer and director of *The Lighthouse*, Charles Dickens was also the lead player under the stage-name Mr Crummles – doubtless inspired by the amusing, grandiloquent actor-manager of that name in Nicholas Nickleby. Also in the cast was the author, Wilkie Collins; the fashionable painter, Augustus Egg R.A.; Dickens's sister-in-law, Georgina Hogarth; Mark Lemon, magazine editor and playwright who, with Henry Mayhew founded Punch; and 'Miss Dickens'. The scenery was painted - naturally enough - by 'Mr Stanfield, R.A.'

There is no record of what the Wardells, favoured friends invited on that occasion, thought of the production, but Dickens's son tells us that his father, as Aaron Gurnock, the head lighthouse-keeper, had an 'extraordinary melodramatic intensity and force'. Ackroyd writes that Dickens noticed with satisfaction afterwards that the audience, limited to twenty-five persons including the Wardells, was 'crying vigorously.'[19]

The infrequent but intense Dickens productions made up only a small part of the Wardells' dramatic entertainments in their Hampstead years. The neighbouring families at The Green Hill had their own less formal theatrical evenings involving children, family friends and neighbours taking part. Some of these 'Charades' as they were called, involved rehearsal periods, printed programmes and 'dressing-up' and were staged at the grandly named 'Theatre Royal', The Green Hill. Whether this makeshift room was in the Stanfields' or the Wardells' side of the terrace is not clear. One programme which survives announces a charade in three acts in which Frank Stanfield plays the piano and George Stanfield plays a Travelling Artist. The cast list includes a 'Mr Wilkinson' billed in one charade as Sir James Compton, a Patron of Art. It is tempting to think this character, in the custom of other pseudonyms used, is Wardell himself. In the second charade 'Wilkinson' is William Tell. Wardell's eldest daughter Mary Wardell, about ten years old at the time, plays the child impersonating William Tell's son who has an apple placed on his head.

The high spirited, humorous, convivial evening ended with an epilogue especially written for the occasion. Because the one that survives is in Wardell's handwriting it seems likely that he was the author. It concludes:

Excuse the actors faulty, and play no critics parts,
Glad to be pleased, though we have no machinery
Nor (though Stanfield's here) the smallest bit of scenery;
Nor can we boast of footlights, lights or rouge.
Borders and flats we equally disuse.
No Manager, no manageress, in short no nothing
Not even a friendly editor nor puffing!
Our Little show is over. Its object will excuse
Its many follies, for we've sought but to amuse![20]

The final item on the programme printed in bold black letters in place of the traditional 'God Save the Queen', was 'God Save the Pope' followed by meaningful exclamation marks. A reminder that this company on that evening was almost exclusively made up of Catholics proud of their religion, and yet still facing an unwelcome amount of ridicule and discrimination.

CHAPTER 7

EMERGING FROM PUGIN'S SHADOW

By mid century, when Wardell's London practice had been established less than five years, Pugin's fame was at its height - although notoriety might be a more accurate word to describe it.

That 'fame,' as Rosemary Hill observes, 'seemed only [to make] him a loss or a laughing stock.'[1] The satirical magazine *Punch* had regularly lampooned him, dwelling on "Pugsby's' Gothicism, his legendary speed and versatility. A cartoon contained a mock advertisement which advertised: 'Cathedrals made in five and forty minutes'[2] The fun even extended, as we have noted, to an amusing reference to him in one of Browning's longer poems.

But behind the fun was the reality of deteriorating health, depression, 'loneliness and wild restless anxiety.'[3] So furiously had his pace of work continued throughout his short life that, by the age of thirty-seven in 1850, he was prematurely worn out both physically and mentally. He wrote to his friend and business partner, John Hardman, 'thus comes the last stage of all sans teeth, sans eyes, sans everything I am very bad tonight...I shall never be well again ... I am done for ...'[4] Even allowing for Pugin's tendency to excessive self-dramatisation, it is clear that by mid century his powers were waning. He remained fully occupied with work, but his Herculean output had more to do with designing furnishings for churches and architectural fittings than working on major commissions to design churches which had brought him fame and relative wealth. 'I am reduced to a mere mechanick a maker of handles to other men's buildings,'[5] he complained.

As Pugin's popularity as an architect of churches declined, so Wardell's popularity grew. As previously argued, in some sense he became a rival to Pugin. We have noted above how Dr Rock, Shrewsbury's chaplain and a close friend of them both - but later to turn against Pugin - wrote to the Earl recommending that Wardell be appointed to finish the work at Alton Towers in place of Pugin. But Shrewsbury remained faithful to his architect in spite of his illness and slackening interest. 'Pugin and Powell hope to carry on the business,'[6] he replied optimistically.

During the eight years left to him before he emigrated, Wardell received over twenty major commissions. By mid century he was held in the highest esteem by leading clergy – those who were in a position of patronage. Later he would be described as 'the ablest of the Roman Catholic architects working in the tradition of Pugin'.[7]

Throughout 1850 and the early months of the following year Pugin, in spite of no new commissions to build churches, was nevertheless fully occupied, still dangerously overworked and under pressure. Dashing frenetically between Ramsgate and central London he was engaged in the design of furnishings for the interior of the Houses of Lords and Commons; and preparing his exhibit for the forthcoming Great Exhibition. The former work was hidden from the public and the latter brought praise and appreciation from the highest in the land. Both the Prime Minister, Lord John Russell (no friend of the Catholic Church), and Queen Victoria spent time talking to him and praising his work. Although the Exhibition was seen and admired by many thousands of people – over six million visitors were estimated to have passed through the doors in five months - and Pugin's name was 'a household word,' it was his reputation as a designer of mediaeval furnishings that attracted this vast public attention, not his reputation as an architect of churches and cathedrals as he would have preferred. After the opening of the Great Exhibition the *Illustrated London News* would gush: 'To Mr Pugin...who furnished the design for this gorgeous combination, is the highest honour due; and he has marvellously fulfilled his own intention of demonstrating the applicability of Medieval art in all its richness and variety to the uses of the present day.'[8]

The shift in public perception of Pugin, and his mental illness leading to his death in 1852 soon after the Great Exhibition, to some extent facilitated the rise to prominence of Wardell as an alternative Catholic architect. But it should not be thought from this that Wardell was not worthy, by right, to take Pugin's place. There were other Catholic architects influenced by Pugin at that time. Among them were J. A. Hansom, who was especially successful (many of his designs were used throughout Australia) and William Goldie, to whom Wardell would eventually sell his practice. Goldie completed some of Wardell's churches after Wardell departed for Australia. But it is arguable that neither of these possessed the level of creative talent, originality, and historicism to match Wardell's. As leading Catholic architects, Hansom and Goldie reigned in England without rivals for a far longer period than Wardell. But there is little doubt that Wardell would have taken his rightful place in the pantheon of England's major, influential architects, had he not disappeared from the English scene in 1858. 'In the mastery of a certain form of decorated Gothic work, in the adaptation to modern needs [of the] true principles of medieval work Mr Wardell was without a rival.'[9]

Just twelve months after the great exhibition closed, Pugin, rescued from an asylum by his wife and close friends, died at his home in Ramsgate on 14 September 1852. His death was overshadowed in the public consciousness by the death of the Duke of Wellington on the same day. Seven days later a solemn Requiem was held in his Ramsgate church, St Augustine's, which he had designed and in great part had financed himself. Among the dignitaries present for the Mass and burial were his friends, family, professional colleagues, and architects who had been much influenced by him. It is most likely that Wardell, a life-long admirer, would have been among their number.

In a sense, Pugin did not die. His mighty influence lived on in the works of his disciples, and particularly in the work of Wardell. Wardell's work embodied Pugin's principles but did not slavishly copy them. 'Every building that is treated naturally, without disguise or concealment, cannot fail to look well' Pugin had written, and 'any attempt to deceive in a building is architecturally deplorable.'[10] These and other ideas 'for which Pugin deserves our gratitude'[11] would inform all of Wardell's architecture.

The eight years that remained before his surprise emigration brought Wardell a spate of commissions which necessitated frequent journeys to various parts of the country. By mid-century he could undertake these expeditions by train, perhaps along tracks which he himself had first surveyed in his training years.

One of his smaller churches about this time was Our Lady and St Charles Borromeo at Wisbech in Cambridgeshire. Wardell travelled there by train via Kings Lynn in Norfolk in the summer of 1854 and supervised its construction. In a letter to Mr Bennett, his builder, he warned of his imminent arrival for an inspection and he shows how exacting in the matter of construction and decoration he had already become, and the demands he made on his contractors: '... the pier of the chancel arch to which I have already called your attention is, I understand, still giving way. I need not remind you of your heavy responsibilities by the contract of taking immediate measures to remedy [the problem].'[12] Two weeks later he was concerned with Mr Bennett's arrangements about the carvings: 'It is a point upon which I am particular and upon which I am obliged to insist.' He tells Mr Bennett that he is unable to make drawings for a forthcoming exhibition in Kings Lynn because 'I am suffering from the effects of an accident and my sight is still very weak'.[13]

Wardell's concern for his Wisbech church and his insistence on standards being met add poignancy as well as disappointment when we witness the inappropriate changes that were made to the building in 1962. A square tower block, solid and rather squat, has been attached to the west end fronting Queen's Road which resembles in style a military bunker. The narrow, elongated slits near the top on three sides of the tower add to the impression of gun emplacements behind. The tower block itself sits on a base of three solid-looking square blocks, brick built, and in an attempt to marry them with the style of the church behind, the gothic, traceried windows of the Wardell section have been reproduced here. The effect is one of even greater incongruity. The original side porch with steep gable has been retained but no longer used. The entrance today is below the tower. Once inside, under the choir gallery, there are familiar Wardell features: five arched bays and aisles either side of the nave are lit by the traceried windows and oriel windows above. The rood beam and crucifix have been retained and happily there is a sense that the church - whatever the mistakes of the ugly sixties – is loved and well-cared for within. The large, painted wooden statue of the Virgin and Child is especially fine and is obviously prized. The tracery of the three-light east window is, unusually for Wardell, curvilinear.

In contrast with the modest size of the Wisbech church, his plan for the

East End church of St Mary and St Michael, Commercial Road, was vast. Wardell started work on the plans in 1852 and the foundation stone was laid by Cardinal Wiseman on May 24 the following year. So imposing is its presence, and so exceptionally long and tall is the roof-line, that it stands out quite distinctly along the busy Commercial Road. Even without the intended but never-realised tower and spire, it has earned the epithet, 'The Cathedral of the East End.' The vibrant life of present-day St Mary and St Michael's - it recently celebrated its 150th anniversary with great pomp - belies its turbulent history and the struggles of the successive founding priests to rally support and raise sufficient funds to pay the builders. In 1851 the district had two inadequate Mass centres, one in Virginia Street and the school chapel in Johnson Street. A new church to accommodate the rising, mainly Irish population was clearly needed. A committee was formed to raise funds and William Wardell was commissioned as architect. His brief was for a church of 'Cathedral-like proportions' with tower and spire soaring to a height of 250ft. Also included would be a spacious presbytery, sacristies, and a lodge for a caretaker.[14] The committee was fired with enthusiasm; money seemed not to be a restraining consideration at that stage. The designs were submitted, tenders were called, and the lowest, £14,674, was selected.

Things got off to bad start with the laying of the foundation stone by Cardinal Wiseman. Mass at a temporary (outdoor) altar was cancelled due to high winds. The Cardinal was unwell and couldn't stay for the reception, and few people attended. Wardell surely would have been one of the few but if he was there, other important guests were not. The hoped-for donations were not forthcoming. Worse was to follow.

Building commenced and the walls reached a height of about 8ft. By that time, however, the early enthusiasm of the committee and the resolve of the congregation to raise money evaporated. The sums collected were piteously small so that the builders could not be paid. They removed their scaffolding and departed leaving a forlorn ruin and the parish in debt.

To the rescue came a remarkable priest who almost single-handily, set about raising money and so well succeeded in organising the affairs of the large parish, that the builders were persuaded to return. Work re-commenced sometime later. The miracle-worker was Father William Kelly, an imposing, courageous and popular man who had suffered various illnesses in the course of his onerous duties in the putrid dockland areas. His sufferings resulted in his being ordered out of London some years previously for the sake of his health. When the new church building project failed and 'morale was very low' the cry went up that only Father Kelly would be capable of solving the problems confronting St Mary and St. Michaels. Efforts were made to bring him back from convalescence in Portsea on the south coast. Understandably, at first he was reluctant to move.

Kelly himself recounts in his diary why he changed his mind. In a vivid dream he heard a voice lecturing him, 'This place is too comfortable for you. Much more is expected of you ... Get away and build the church for those poor people in Virginia Street district.'[15] He recounts that for one whole week his nightly sleep was disturbed in this disquieting fashion.

Father Kelly put aside his hesitation and his health concerns, and with the Bishop's approval returned to his old London district on 18 October, 1853. He noted in his diary that it was a cold cheerless day: 'I found myself in the Commercial Road, East, viewing what might have been the ruins of some large buildings destroyed as if by an earthquake, with some low stunted pieces of walls standing here and there upon them. Not a bit of scaffolding was to be seen in any direction. The builders, as I learned subsequently, had a claim of £500 upon this seeming wreck. My predecessor had paid them all he had in hand shortly before my arrival, so that he had not even sixpence to hand over to me to meet so large a balance.'[16]

He tells us he was tempted to listen to those who advised abandoning the idea of such a large and expensive church, but he resisted the criticism, 'To some it seemed like insanity to continue the building of the church, and so Job's comforters gathered around and earnestly urged me to pull up the foundations, and with the brick and stone erect a shanty here and a few others in different parts of the parish.'[17] Fortunately Job's comforters did not manage to convince the priest. He admitted that it may have been a mistake to commission so large a church but he felt it was too late to withdraw. He believed he had a duty to provide adequate Mass facilities for the growing East End population.

Father Kelly's strategy was to conduct a census in the surrounding district and sign up each Catholic for a monthly financial contribution according to the individual's ability to pay. Then he assigned collectors to collect the money. He describes his remarkable work in his journal, 'The time daily employed in making the census ranged from 4½ to 9 hours ... it may be stated, free from exaggeration, that for six days out of seven every week, five hours daily during a period of five months and fourteen days, were spent in this work of census making, going personally from house to house and penetrating from cellar to garret.'[18] Painstakingly he went off each day 'large book in hand, and pen and ink in waistcoat pocket' going from house to house, often room to room because in the miserable tenements each room might accommodate a different household. Adding to his troubles were debts on the three schools in the district and the forced closure of one of them. 'Such work as this I never expected. The children were ejected,'[19] he notes. Also, during his work on the census a cholera epidemic was raging violently in the district. Kelly describes how he was frequently called back to a house he had visited a few hours previously on his rounds in order to administer the sacraments to a dying occupant. When completed, Kelly's detailed statistics revealed the Catholic population of his parish district to be sixteen thousand of which total five thousand two hundred and forty-eight were children under 14 years. The capacity of each adult to contribute was assessed but in most cases it must have been a pittance. The daily rate for work in the docks ranged from one shilling to two shillings and sixpence. Work was irregular and affected by the weather. Thousands of waterside workers were thrown out of work in the winter time. In 1855 the Thames froze over and prevented ships from docking – the worst blockage for forty years. Workhouses distributed outdoor relief, contrary to policy. Kelly noted in his diary: 'Thus it is that the deepest direst distress pervades our poor amid this awful inclement season...The small offerings of our poor boxes are gone – our own scanty purses are empty in their relief, and still the cry in our ears is, bread, bread.'[20]

Before the 1855 winter conditions wrought their misery, Kelly had collected sufficient money to persuade the builders to return to the church site. Summer weather had been a time of employment and the flow of contributions and a modest grant from the Catholic Parishes

Support Committee (CPSC) had resulted in the tower of St Mary and St Michael's rising to a height of forty feet. But with the approach of winter Father Kelly knew progress would slacken. He wrestled with fresh difficulties and had to make new economies. By August 1856 he still needed £5000 pounds to complete the building but 'there was not one farthing of balance in hand to pay for the next monthly instalment.' To save time and money he decided the Wardell's spire would have to wait for better times. The truncated tower which we see today was then roofed over with a gable at a height of sixty feet.

This brief outline of the torturous story of the building of St Mary and St Michael's is given to show how close Wardell's largest London church came to being abandoned at an early stage. Now, as visitors and the contemporary population look at the imposing exterior in the Commercial Road, a landmark in the East End, the struggle, the suffering and the ill-afforded pennies of the mid 19th century faithful are as distant, unknowable, and as misunderstood as were the faith and generosity of the mediaeval builders of Chartres and the great cathedrals of Europe.

For those who step inside the church - and for fear of vandals it is generally closed between masses - the effect is magical. The vista is remarkable for its openness and light, and for its great length and width (180ft by 75ft) – unusual for what is, after all, a parish church in a poor area of London. The perspective, uninterrupted by transepts, arches or screens, is breath-taking. Here we see Wardell's particular genius for harmonising space and mass and imprinting his own character and delicacy on Decorated Gothic. Eleven continuous arched bays run the entire length and light floods in from the windows of the aisles and the clerestory above. The whole floor retains the original red and blue encaustic Staffordshire tiles. A greater surprise perhaps, is to turn and face the entrance and admire the rich mouldings of the giant arch which is all that is left to mark the ground floor level of the tower and steeple. The five-light window at this point is modern and so too is the high altar and reredos at the south end - the church lies on a north south axis.

The exterior of St Mary and St Michael's is in Wardell's favoured Kentish ragstone with Caen stone dressings. Seated on top of the buttresses, either side of the north window, are statues of St Michael and St Mary, and in a niche in the gable a statue of Christ the King. Unhappily the roof was severely damaged by bombing in the area during the war and required extensive repairs. The massive area covered by smooth, modern slating tends to sit uncomfortably on the weathered stone walls.

St Mary and St Michael's is a church to visit, to admire and to remember in prayer the sacrifices and sufferings of the struggling poor who, led by their heroic priest Father Kelly, succeeded in realising Wardell's vision.

Wardell's order book filled rapidly in that first decade of the second half of the century. Among commissions are listed designs for a church and school in Manchester; St Richard's in Chichester (now demolished); the Church of St Mary and Presbytery in Chislehurst in Kent; a wing of the Convent of the Faithful Virgin, Upper Norwood; the Church of Our Lady and St Joseph in Poplar which - like St John the Baptist in Hackney - was destroyed during the war. With these and others, Wardell was now firmly established in his profession and not lacking in ambition. He sought status and on payment of a fee was made a Freeman of the City of London – a nominal honour which was considered an aid to professional advancement.

Often absent from home, his work took him to all parts of the country. He was certainly absent from The Green Hill on 30 March, 1851 because he does not appear in the census on that date, neither in Hampstead nor elsewhere in England – which suggests that he may have been making one of his periodic visits to the Continent. Absent on the Continent he may have been at the end of March 1851, but he was certainly there in July of that year carrying a letter of introduction from Father de Hels of Clapham addressed to his Superior in Paris introducing 'Monsieur W. W. Wardell, architect, qui a bati notre église à Londres'... C'est un des nombres beaux converts en notre époque.'[21] Wardell was evidently well known in France among French antiquarians – friends of Pugin's – among them, eminent architect and theorist, Eugène Viollet-le-Duc, and le Comte de Montalembert both of whom led a successful campaign for the restoration of historic monuments of France. Undoubtedly it was his association with these two that led him to seek nomination and be accepted as a member of the *Societé Française pour la Conservation et la Description des Monuments historiques*. In his citation from the Society, dated 20 February, 1855, Wardell is charged with doing all he can to further the ideals of the Society in his own country and 'prevent the degradation of historic buildings and save those whose existence is threatened.'[22]

His absences were not as burdensome for the family as they might appear to us today. Lucy Wardell employed two live-in servants, a housekeeper, plus a nurse for the two children. The eldest girl, Mary Lucy, was about three, and Edward aged one at that time. And there were always the Stanfields next door, including six children and four servants to help with them. That the two families were in and out of each others' houses is clear from a remark in a letter written by Stanfield to Wardell after the Wardells had arrived in Australia. There he admits that he had not met the new occupants next door because it would 'break his heart to enter the old familiar rooms'.[23]

We know from other correspondence that Wardell was back at The Green Hill certainly by June of 1851 when he surely would have regarded a visit - or multiple visits - to the Great Exhibition in Hyde Park not only as entertainment for the family but for him, professionally instructive. Hundreds of thousands of people of all classes across England took advantage of cheap train travel and joined the many organised excursions to visit and admire the great Palace of Glass in Hyde Park, and marvel at the exhibits. As an engineer Wardell would have wanted to study them in the machine room and elsewhere:

> 'It is a wonderful place – vast, strange, new and impossible to describe. Its grandeur does not consist of one thing, but in the unique assemblage of *all* things. Whatever human industry has created you find there, from the great compartments filled with railways, engines and boilers, with mill machinery in full work, with splendid carriages of all kinds, with harness of every description, to the glass-covered and velvet spread stands loaded with the most gorgeous work of the goldsmith and silversmith, and the carefully guarded caskets of real diamonds worth hundreds of thousands of pounds.'[24]

The 'gorgeous work of the goldsmith and silversmith' reminds us that Wardell would have had a personal as well as professional reason for visiting the Exhibition. He would have wanted to see his friend Pugin's last work, his creation of the most popular and splendid Mediaeval Court exhibit, much admired by Queen Victoria and Prince Albert, taking up a prominent position in Paxton's building of glass. The *Illustrated London News* judged it the best thing in the Exhibition.[25]

This was to be Pugin's triumphant exit from the public stage. The Crystal Palace closed its doors in the first week of October 1851 and Pugin found himself honoured by an appointment to the powerful committee that decided which of the exhibits should be bought for the nation – for housing in the future Victoria and Albert Museum.

But within twelve months Pugin, the most influential architect and designer of the Gothic Revival - and possibly of the 19th century - would have left the stage, and William Wardell was emerging from the wings.

CHAPTER 8

A CHANGE OF CLIMATE

Early in 1858 the Wardells awoke early one Spring morning to face the prospect of a complete change in the direction of their lives and the severance of their stimulating social connections. And the death in March of their third son, Michael Thomas, at the age of five may have had something to do with it.

The family might have hoped that they were relatively safe from the fetid air and disease-bearing fogs which blanketed the city located to the south of them, but this was not so. When viewed from their high vantage point of Hampstead Heath, London was invariably blanketed by fog, a filthy city, plague-ridden and smelly. In Hampstead, a northern suburb, the air was clearer and supposedly healthier but although the village benefitted from the high ground and open Heath nearby, the district could not escape entirely the notoriously unhealthy city conditions. Wardell himself travelled into that fetid city, to his office each day and seems to have believed that his health was adversely affected. (Thomas Hardy, another young architect who arrived in London four years after Wardell had emigrated, described in a letter to his sister how 'swarthy columns of smoke rose from the massed kitchen chimneys every morning, spreading out to form a haze that darkened the sun and gave the air its city smell.'[1] He did not stay long before returning to Dorset).

We do not know precisely when William Wardell realised that the climate was affecting his and his family's health, or when he himself was ill enough to consult his doctor. Nor do we know what symptoms convinced him to do so. But we can guess, in view of the advice he was given, that he was experiencing a persistent cough, pulmonary discomfort, perhaps a trace of blood in his sputum, and a growing listlessness. Before the wide-spread use of the stethoscope,[2] and long before radiography, doctors relied to a large extent on the patient's appearance, and their precision in describing their symptoms that would lead to a diagnosis of tuberculosis. In Wardell's case the disease must have been either at a very low level or, what is more likely, was misdiagnosed altogether – a common enough error in those primitive medical years. As there is no evidence of a recurrence of the disease in later life - a life lived to a comparatively old age – the diagnosis remains at best, speculative. In hindsight we may doubt it, but the doctor's advice was believed by Wardell in 1858, and his condition serious enough to convince him that he should give up his home and work in London, and journey to a better climate elsewhere. The city was a notorious incubator of disease, and those sufferers who could afford to do so would seek a more beneficial climate elsewhere. Had not Keats (a resident of Hampstead), Ruskin, and Elizabeth Barrett Browning and many others, done this before him?

At that time, mid-century, tuberculosis was 'the single worst disease cultivated by monster cities, invariably a killer believed to be responsible for an average 50,000 deaths in London each year – more than smallpox, typhoid, scarlet fever, measles and whooping cough put together.[3] There was no sure cure; many sufferers meekly accepted their fate, and others sought well-meaning but ineffectual remedies. Admirers of the Brontes will recall that Charlotte described her sister Emily as having 'a hollow, wasted, pallid, aspect' and that 'she resolutely refuses to see a doctor'.[4]

Wardell, in accepting the validity of the doctor's diagnosis was convinced of the gravity of his situation. Nothing less than tuberculosis, surely, would have induced him to give up his successful practice in London and his valuable contacts, and migrate with his family to a more equitable climate in a strange colony on the other side of the world.

We can only surmise what anguish Lucy Anne suffered at the prospect of parting from her parental family in nearby Oxford, and from her friends including the Dickens family and the Stanfields next door, and so many other writers and artists in their circle. Since migration in slow sailing ships in those days was not only hazardous but generally irreversible. An inevitable break in those friendships - the family's social evenings, their shared intimacies, and their outings together - must have hurt her beyond our imagining. But in Victorian London a wife deferred to her husband accepting this as her duty. She would have believed that he, as head of the household, knew best what to do, probably without question or objection from other members of the family.

But were the reasons for sudden emigration more complex than health? In Victorian times - as now - there resided, embedded deep in the human spirit, a restlessness which, in vast numbers of people, could only be assuaged by travel and even emigration. Although the widely accepted reason for William Wardell's emigration has been explained by his deteriorating health, perhaps the motivation was more complex. This was hinted at in a speech by Bishop Morris at a farewell dinner for Wardell at Greenwich. The Bishop expressed his regret 'at his [Wardell] being compelled, *principally on account of his health*, to quit his native shores.'[5] [italics added] A slip of the tongue? Or does the careful wording hint at more than one reason for Wardell's decision to leave? Wardell the architect was an ambitious man and would have been aware that many in his profession who had talent and a business sense had risen in status and amassed considerable wealth - Sir Robert Taylor, Sir William Chambers, Henry Holland and Sir John Soane to name a few of his contemporaries.[6] And if, as we suspect, Wardell had met visiting Archbishop Polding, Bishops Goold and Willson and they had spoken to him of the need and the opportunities for an experienced architect in their Australian dioceses supported by the new-found wealth of the Colonies, the idea of emigration might have been simmering in his mind before the question of his health had arisen.

The children probably viewed the changes as an adventure: Edward, the eldest boy aged nine remained a border at Stonyhurst Preparatory School and Mary Lucy, aged ten, was then boarding at a convent in Sussex. Francis William (2), Kathleen Mary (3) and Ethel Mary (1 year) would remain behind and sail with a nurse on a later voyage of the *Swiftsure*. Only Bernard, aged six, would accompany his parents.

Why Melbourne was chosen as the preferred destination for a tuberculosis sufferer is not clear. A climate considered beneficial would have been an important factor, but it seems Wardell was almost certainly advised of the Church's needs there and the dearth of good architects and engineers in a town that was expanding rapidly on the back of the rich gold discoveries in the area. It is significant that in January of that year Wardell first applied, and was elected, an Associate Member of the Institution of Civil Engineers. He gives his business address at the time as 44 Parliament Street, Westminster, within sight of the new Houses of Parliament recently completed following the disastrous fire of 1834.

In the first months of 1858 Wardell set about the arrangements necessary for departure and, characteristically, attended to every detail with as much care as he was in the habit of giving to the design of one of his buildings.

In those remaining six months of 1858 a number of commissions for new churches demanded his attention. Among them were two churches in the Border country of Scotland – Our Lady and St Andrew's in Galashiels, an uncompleted part opened in February of that year; and St Mary of the Immaculate Conception in Kelso, which was opened while Wardell was on the ship heading for Australia. Both commissions had arisen from Wardell's long-standing friendship with Robert Hope-Scott, barrister and Member of Parliament. Hope-Scott's wife, Charlotte, was the granddaughter of Sir Walter Scott and the couple had inherited Abbotsford House, the author's romantic Gothic Revival mansion on the banks of the Tweed. They also kept a London house opposite Hyde Park near Marble Arch where a frequent visitor was the then Anglican priest John Henry Newman. Newman was embroiled at that time in the controversies surrounding what was later to be described as the Oxford Movement. After Newman became a Catholic the Hope-Scotts followed him into the Church – James Hope-Scott being received into the Church alongside his close friend, John Henry Manning - later Cardinal Manning - at the Jesuit church in Farm Street on Passion Sunday, 6 April 1851. Thus two more prominent former Tracterians joined that fascinating coterie of convert professionals to which Wardell and Pugin, Newman, Stanfield, Shrewsbury and Ambrose March-Phillipps, and many others belonged.

The first extant letter from Hope-Scott to Wardell is dated Abbotsford, 6 February 1850, 'I am very glad to have such good news of your children,'[7] he wrote, and then he refers to land in Galashiels purchased by him for a chapel already built there. With the increase in manufacture of Tweed cloth at Galashiels and the construction of the railway, the population grew significantly. The chapel soon became inadequate, and with great generosity and zeal Hope-Scott bought more land adjacent to the chapel with the intention of building a large new church situated close to the railway station - now the central bus station. Late in 1856 Wardell was asked to design it, and construction began soon afterwards.

When visiting Galashiels, Wardell stayed with the Hope-Scotts at Abbotsford House nearby, and would have attended Mass in their private chapel – an innovation hardly likely to have pleased the solidly protestant Sir Walter had he been alive at the time. The tastefully furnished little chapel still remains, to the fascination of tourists, and Mass is celebrated there occasionally. It has a mediaeval stone fire-place designed by Wardell in keeping with the style of the neo-Gothic mansion.

Today, the imposing Church of Our Lady and St Andrew in Galashiels overlooks the town from its exposed position on high ground, and is much admired. The size reminds us of St Mary and St Michael's in the Commercial Road although its dimensions are slightly smaller. Like the East End church, its Scottish cousin also presents a massive aspect due in part to its one hundred and twenty-five feet length and steep-pitched slate roof seventy feet high at its apex. And again, like the London church, it lacks a tower or spire. But there the similarity ends. The exterior is of local russet-tinted whinstone, brooding over the town in dull light but seeming to glow in the sunshine. The west front is flanked either side by five-sided pilasters - small towers - which reach up to the top of the gable and are capped by quintagonal turrets. This, together with the seven great buttresses with pedimented heads puts us in mind of Kings College, Cambridge. The resemblance, however, is only a fleeting one. The great five-light window of the west front has curvilinear tracery and above it is a small round window in the gable. Entrance is by one of two small doors right and left of arched, blind arcading at street level.

Inside, the first impression is of the massiveness of construction, and instantly we are struck by the height which soars upward to the great timbers of the roof and appears out of proportion to the width. An unusual feature is the seven bays either side of the nave formed by the exterior buttresses being brought within the walls to form a series of chapels, confessionals, and a sacristy. At the east end, above the High Altar is a pair of three-light windows and a spherical window above to match the one in the west gable.

The great church was still under construction when the Wardells departed for Australia and the decorative features were supervised by Hope-Scott himself. In a letter to his friend the recently ordained Father

John Newman - later Cardinal Newman - he wrote:

> 'I shall not be hurried in the decorative part, which I cannot afford to do handsomely at present, and which I think will be done better when we have become used to the interior, and observed what is to be brought out, and what is to be concealed. The shell I am well pleased with. It is massive and lofty, no side aisles but chapels between the buttresses - and no altar screen – more like a good college chapel than a parish church....to keep us in mind that more is to do, we have a rough temporary work at the west end, with square sash windows of the repulsive aspect.'[8]

The completion and official opening would not occur until five years after Wardell's departure; the finishing stages were supervised by George Goldie. Hope-Scott did not live to see the completion of his Galashiels church, but died prematurely three months before the opening.

However, he surely would have been present for the ceremonies on 8 September, 1858 when the parish church of the Immaculate Conception was formally opened and dedicated in nearby Kelso. This was another Wardell design commissioned and partly paid for by Hope-Scott. Although much smaller and less ambitious than Galashiels in keeping with the small town and population, it stands today on the site of the first chapel and Catholic school in the district. The original buildings were gutted in a fire which was ignited by anti-Catholic rioters on the night of 6 August, 1856. Wardell's plans show the building which replaced it in the form of a cross, the transepts were to consist of a Lady Chapel on the south side and a sacristy on the opposite side. Funds were insufficient to complete the project. The nave was modified very simply and the chancel was first built in brick being regarded as only temporary. The chancel as it stands today was designed by Archibald McPherson, a local architect, and incorporates stained glass windows by Hardman. The completion had to wait until 1914-15.

The well proportioned exterior is especially pleasing in local stone, the effect belying the very plainness of the nave inside. Already we notice the Wardell Gothic touches, characteristic of his smaller churches both in England and Australia - the steep roof and gable end, and the bellcote rising above, all in local stone. Viewed from the outside it is clear that the church was intended to be of a more complex design inside than now appears.

As the months leading to Wardell's departure wore on, his order book was well-filled with commissions and we do not know how many offers he then had to refuse. Chief among those that claimed his attention at that time would have been the church and presbytery of St Mary & St Edmund's in Abingdon near Oxford, and the church and presbytery of St Mary's in Axminster, Devon.

Wardell's first plan for the Abingdon church is dated 1856 and addressed to Sir George Bowyer M.P. who, like Wardell, was a convert Catholic. Bowyer, who would later write a glowing testimonial for Wardell, was a leading figure in the Catholic 'Second Spring,' a friend of Cardinal Wiseman and the main benefactor of St Mary and St Edmund's. The grand opening ceremonies of the church took place on September 30, 1857 nine months before Wardell sailed for Melbourne. He and his wife would surely have been among the distinguished guests - many of whom from old recusant families: those who had retained their Catholic faith during centuries of persecution. *The Tablet* reported that 'many had travelled some distance and having endured the lengthy ceremonials sat down to a banquet in the cloister. The numbers requiring lunch – some one hundred and fifty – obliged the organisers to arrange two sittings, the ladies first, followed by the men.'[9]

The church was then far from complete and would have looked very different from its appearance today. The interior nowadays is disappointing. Wardell's chancel on its North East axis has been 'reordered', that is, stripped of its original furnishings and decoration to reflect modernist principles in the early 1970s. A parishioner who came to the parish in 1974 reports that few seemed happy with the changes at the time, and within a few years many of the key features were 'quietly dismantled'. The cost of the reordering was in the region of £25,000.[10]

The great stained glass window with its five lights and tracery above – a fine example of Wardell's many window designs - relieves the plainness and anonymity of the sanctuary. Fortunately Wardell's south aisle chapel which contains a statue of St Edmund has retained the original decoration and furnishing inspired by Pugin, giving us a taste of how the entire church might have appeared had it been completed as envisaged by its architect. The nave was completed by George Goldie after Wardell's departure but is said to be true to Wardell's original plans.

The different shades of the exterior stonework and the tiles of the roof suggest piecemeal construction. A more pleasing view of the complex can be obtained from the old graveyard on the north side showing Wardell's cloisters connecting the church with the imposing Gothic presbytery. On completion, the church and presbytery must have done much to enhance the Abingdon townscape, as evidenced by a report written by a local journalist who thought it 'the very best of the many successful instances of the glorious Gothic architectural revival of the past few years by the architect Mr W. W. Wardell, of Parliament Street and Hampstead'.[11]

Another church, unfinished at the time of Wardell's departure, is St Mary's and its adjoining presbytery on Lyme Road leading up the hill out of Axminster towards the coast - its full title St Mary's Catholic Church which distinguishes it from the 13th century Minster and Parish church of St. Mary the Virgin in the centre of the town. Here again was an instance of a rich local patron, Henry Knight, purchasing land in 1854 and commissioning Wardell to prepare plans. But there was considerable delay due to the illness and death of the patron. Not until 1858, the year of Wardell's departure, did the building work commence.

The result is a charming country church in local stone complete with a gabled porch in wood and stone. Inside, the unadorned nave is divided from the chancel by an arch, and this is enhanced by a stained-glass window of four lights over the altar at the east end. As the sun comes up in the mornings it streams through the coloured glass and illuminates the sanctuary giving an ethereal effect. Like Abingdon, St Mary's Axminster was also completed by George Goldie faithful to Wardell's plans; the total cost including presbytery and adjoining school was reported to be £3000.

Not least of Wardell's concerns in preparation for his departure was the disposal of the lease of the Hampstead house, and the sale of his architectural practice which went to Hatfield and Goldie. He had also to devise a means of establishing a similarly successful practice in Melbourne. He set about this in a typically thorough and ambitious way by writing to some dozen of his most prominent clients and friends seeking their testimonials. Heading the list was Archbishop - later Cardinal - Wiseman of Westminster, Bishop - later Cardinal - Manning; the Bishops of Southwark, Northampton, Troy and Liverpool; and among other leading names, Lord Petre for whom he had built a chantry chapel in Essex, Sir John Simon, George Bowyer M.P. - who financed the Abingdon church - , and Wardell's great friend and neighbour, the painter, Clarkson Stanfield. The latter wrote: 'few are better qualified than myself to speak of your private worth in your domestic circle and the appreciation in which your talent is held by the general public and your brother artists.'[13] All wrote glowingly of his professional competence, his achievements, and his virtuous character. Wardell then had these testimonials printed in pamphlet form by Waterlow and Sons and copies were distributed in advance of his arrival to all prospective clients in Victoria, including all the clergy there. This self-promotion on such a scale may seem to us somewhat overblown, but indicates Wardell's level of ambition, and his confidence in his professional self-worth. The CEO of today, or a university graduate, might employ much the same strategy.

There is no doubt from the evidence available that Wardell's imminent departure for Australia was received by his friends and his clients first with surprise, and then with genuine regret and great sadness in all cases. *The Tablet* fully reported a farewell dinner given in his honour at the Trafalgar Hotel, Greenwich, on 17 June. The Bishop of Troy, Dr. Morris, presided and in 'a touching and elegant speech gave expression to the feelings of deep regret of the friends of Mr Wardell.' The Bishop wished him every success which his brilliant talents deserved. Wardell returned thanks 'for the kindness of his friends, and for expressions so flattering to him, and so consoling under the circumstances.'[13] Other speeches followed highly complimenting Mr Wardell upon 'the splendid specimens of Gothic Architecture which would carry his name down to posterity in this country and would tend to illustrate and glorify the Catholic Religion for centuries to come.'[14]

The *Weekly Register* also announced with regret the imminent departure of Wardell 'who ranks among the first of living architects of England, and than whom no man is more universally respected and beloved.'[15]

As important as Wardell became in the history of Australia's architecture, it is tempting to speculate what mark he might have left on England's architectural history had he remained there and, we may presume, continued to rise in fame and stature. England tends to forget -

even ignore - her sons of great talent who leave to work overseas no matter that their work overseas might compare more than favourably with work done by others who remain in the mother country. Such was the fate awaiting Wardell who, if he be mentioned at all, now occupies merely a footnote or a small paragraph in English histories of architecture.

His lease on his home, The Green Hill, in Hampstead, either expired or was disposed of; his practice sold, and his commissions - even if not all completed – were placed in the safe hands of another Gothic Revival architect, George Goldie. William and Lucy Wardell, together with one child, then packed up, minus the rest of the growing family and boarded the one hundred and thirty-two ton sailing ship *Swiftsure* in the West India Dock. Almost within sight of his boyhood home the ship slipped her moorings on 2 July, 1858 and headed down river to the Channel and south west to Plymouth - her official port of departure for Australia. Curiously, a mistake in the *Swiftsure's* passenger list describes William Wardell as a solicitor.

No stranger to sea voyages, the 'solicitor' who stood on the deck looking at the fading shore-line was then a mature, handsome thirty-five year old – although the waxen image in his photograph, taken only a little time later in Melbourne, hints at his possible ill-health. Nevertheless he manages a determined, serious expression. His receding hairline above a high forehead and side whiskers differs markedly from the youthful likeness sketched by his friend Stanfield. In the Melbourne photograph, in spite of the drawn expression, the eyes, mouth and general appearance suggest a man of fashion, confident and ambitious.

He and his young family would need that confidence as they faced an uncertain future, and certainly an uncomfortable voyage to Melbourne on the other side of the world.

Mary Elizabeth Wardell beside a bust of her husband Thomas Wardell. The only known photograph of William Wardell's mother and father dated 1860. The bust was modelled by William's brother, Herbert Samuel Wardell, in 1850.

All Saints Parish Church, Poplar where William Wardell was baptised and where he worshipped when he was a boy, The church was designed by Charles Hollis in the popular style of Christopher Wren and consecrated in 1823, the year William was born.

The Old Workhouse (and Town Hall) in Poplar High Street, to which Wardell's parents were appointed the Master and Mistress in 1830.

Sketch of 'The Green Hill' terrace by George Clarkson Stanfield. Wardell's house is the double one on the left (now demolished); the one on the right, with studio attached, was the home of the artist, Clarkson Stanfield, and still exists.
Courtesy, Pieter Van De Merwe.

LEFT: Stanfield House, High Street, Hampstead, leased by the artist Clarkson Stanfield. Wardell's house, of similar style, was part of the terrace, the neighbouring house to the left but demolished to make way for Prince Alfred Road.

ABOVE: Parliament Street, Westminster where Wardell had his architectural office. Two Georgian houses survive in the centre giving an idea of Wardell's offices at number forty-four.

LEFT: Wardell's parish church when he lived in Hampstead, St Mary's, Holly Place. Only the Italianate frontage is known to be his design.

RIGHT: The interior of St Mary's showing the additions to the sanctuary by Adrian Gilbert Scott in the 1930s.

Wardell's romantic village church, St Birinus, Dorchester-on-Thames, Oxforshire. Opened in 1849. His earliest existing church in England and best preserved without later refurbishment. The Rood Screen and furnishing owing much to the influence of Pugin.

RIGHT: Church of the Most Holy Trinity, Brook Green, Hammersmith. Opened in July 1851. The Tower and Spire, intended by Wardell, was added by J. A. Hansom in 1871 after Wardell had left for Australia. Evinson considered this the 'High Watermark' among Wardell's London's churches.

ABOVE: Wardell's almshouses adjoining the church and reminiscent of a medieval Oxford College. The view across the green quadrangle, once open to the street, is now blocked by a housing development.

Church of St Mary & St Michael, Commercial Road in the East End of London. Foundation stone laid in 1853 but owing to severe poverty in the district, building was delayed. Part opened in 1856 and became known as 'The Cathedral of the East End' – the largest Catholic parish church in London. Severely damaged during the war and restored. A tower and spire over the entrance was planned by Wardell, but never built.

101

LEFT: Our Ladye Star of the Sea, Greenwich. Opened in December 1851. The first parish priest struggled to find the money to complete it and asked visitors and passers by to 'put their hands in their pockets'. The spire was designed to be visible to sailors in the Thames Estuary.

RIGHT: The nave and angel roof has suffered from clumsy 1960s additions but a remarkable survival is the screen before the chancel in white Caen stone comprising three bays of traceried foiled circles. The church roof was damaged during the war and again in the great gale of 1980.

Church of St Mary, the Immaculate Conception, in the small Scottish town of Kelso. Wardell's first commission to come to him following destruction of an earlier church by fire (The second was St Mary's Cathedral in Sydney). The artistically pleasing exterior in local stone with its characteristic bellcote belies the simple four-square nave and lack of decoration inside. Wardell's plan called for a transept housing a Lady Chapel on one side and sacristy on the other but insufficient funds never allowed completion as intended.

103

Wardell, a superb draftsman; made sketches of his designs for his clients. This one is of St Mary's in Axminster, Devon c1858.

Axminster, Devon. St Mary's Church and presbytery in Lyme Road. Wardell was commissioned to submit designs in 1854 but considerable delay following the death of the patron, Henry Knight, a wealthy Catholic landowner, resulted in St Mary's being completed by George Goldie after the departure of Wardell. ABOVE: The rising sun catches the eastern gable and shines through the stained glass window onto the sanctuary inside. BELOW: A view from the south side showing the back of the presbytery.

ABOVE: Abingdon, Oxfordshire. Church of Our Lady and St Edmund of Canterbury opened, unfinished on 30 September 1857. The nave completed by George Goldie after Wardell's departure. The original chancel has suffered from later 'reordering' in line with modernist principles and liturgy.
BELOW: Also by Wardell is the substantial Gothic Revival presbytery linked to the church by a cloister. One of Wardell's largest and most successful but few domestic buildings.

Church of Our Lady & St Andrew, Galashiels. Partially opened, unfinished, in February 1858 five months before Wardell departed for Australia. Unusual features include the side chapels formed by internal buttresses, and the two separated stained glass windows at the east end. Completed and solemnly consecrated in August 1873 but neither the architect, nor the great benefactor, Robert Hope-Scott, was present. Wardell was in Australia and Hope-Scott had died in April of that year.

Wardell's massive 'Cathedral of the East End' seen from across Commercial Road. The gable replaces the tower which was planned but never built owing to insufficient funds.

SS Mary and Joseph's Catholic Church in Poplar, built to Wardell's designs in 1851–6; now demolished.

William Wardell. A formal photograph dating from the time of his arrival in Melbourne 1858-9. Copyright: MDHC Catholic Archdiocese of Melbourne

MELBOURNE

1858 – 1878

CHAPTER 9

THE NEW SETTLER

The *Swiftsure* entered Port Phillip Bay and anchored in Hobson's Bay off Port Melbourne on Wednesday, 29 September, 1858. The Wardells had been at sea for eighty-four days.

In William Wardell's luggage, in addition to the glowing letters of recommendation and professional certificates, were the plans of all the Gothic churches, schools and presbyteries, that he had designed in England and on which he would base his work in Australia. He had, as yet, no reason to think that his future in the new colony would not be devoted mainly to ecclesiastical building - as we know it had been in England. Like many new arrivals both before and since, he may have thought of his sojourn in Australia as only temporary; that when his health had improved he would return to take up his career again in London. Lending some support for this view was the fact that he had left Edward and Bernard, his eldest sons, at Stonyhurst, the Jesuit college near Preston. They were then aged eight and six respectively,

We should not be surprised that Wardell chose Melbourne for his new home rather than Sydney or elsewhere in Australia. Victoria had been declared an independent self-governing colony seven years prior to Wardell's arrival. Coincidentally with independence in 1851, the discovery of gold at Ballarat had generated not only an unprecedented rise in population so that Melbourne grew from thirty thousand to one hundred thousand in the course of two or three years, but also provoked a general feeling of optimism and civic expansion. This would surely present professional opportunities for a well-qualified architect-engineer. Writers at the time were in no doubt that Melbourne was the capital of Australia, '... the metropolis of the Southern Hemisphere,' enthused the journalist, Richard Twopeny - although he didn't think much of the appearance of the 'capital'. 'The natural beauties of Sydney are worth coming all the way to Australia to see; while the situation of Melbourne is commonplace if not actually ugly; but it is in the Victorian city that the trade and capital, the business and pleasure of Australia [are] chiefly centred.'[1]

Trollope, writing four years after Wardell's arrival, reminds us that forty years previously - only thirty-five years before Wardell's arrival - 'the foot of no white man had trodden the ground,' and so he marvels that 'no city had attained so great a size with such rapidity.'[2] Like Twopeny quoted above, he is critical of the surroundings and the general scenic landscape which he thought 'uninteresting', but he added, 'the internal appearance of the city is certainly magnificent.'[3]

The appearance of the city to Wardell's trained eye - four years before Trollope enthused - must have been more than a little disconcerting; before, that is, he and the Public Works Department in future years had had the opportunity of adding significantly to the built environment. The streets were wide and dusty being gravel surfaced, 'yet no street is finished. Even in Collins Street the houses stand in gaps.'[4] On each side of the main streets ran deep gutters which served the purpose of main drains, and in periods of heavy rain overflowed so that Trollope noted the melancholy fact that children had been drowned therein. At least one sight would have been familiar to the Wardell family accustomed to denser traffic in the London streets and the attendant smells, and that was the presence of several thousand horse-drawn vehicles and the army of council workers whose main occupation was clearing away the horse manure.

Another sight that would have pleased the family on arrival was the warmth and clear blue skies of Melbourne's Spring. Those who live further north in Australia often derive amusement from the changeable weather in the Continent's most southerly city where the winters can be wet and cold and the average rainfall fifty-eight inches. However the climate is in fact generally mild, similar to that of Lisbon, or Washington without the snow. Although the leafy, deciduous trees, such a welcome feature lining the streets of the modern city, were yet to be planted in 1858, the sunshine, the blue skies and the clarity of the air would have heartened the travellers. Shortly after their arrival the Wardells found a temporary home in Powlet Street, East Melbourne. The house no longer exists.

One of the leading figures in the colony whom Wardell would have had on the top of his list of persons to call upon, was the Catholic Bishop (later Archbishop) James Alipius Goold. Wardell would surely have sent to him, ahead of his arrival, his finely-printed testimonials together with a letter offering his services.

Ordained an Augustinian priest in Perugia in the Papal States in 1835, Goold had been consecrated Bishop in Sydney twelve years later by Archbishop John Bede Polding. The ceremony took place in the unfinished first Cathedral of St Mary's designed in part by Pugin and eventually to be replaced by Wardell.

Goold was known to be a well-educated Irishman who had an appreciation of art and architecture. These sensitivities quickly convinced him that St Francis's Church in the heart of Melbourne where he was enthroned, was not suitable either in size or splendour, to be the Cathedral of 'the metropolis of the Southern hemisphere', the self-styled capital of Australia. Goold was determined that Melbourne should have not only a cathedral that would compare favourably with the great cathedrals of Europe, but also be a testimony to what he saw as the pre-eminence of the Catholic Church in the new colony. Until the arrival of Wardell efforts to achieve this aim had been disappointing.

The church authorities had had the advantage of two acres of high ground allotted to them as early as 1848, 'the best ecclesiastical site in Melbourne'.[5] Some believed this was due to favouritism or covert influence, which provoked 'an outbreak of sectarianism which had been almost entirely absent from the earliest days of the settlement ... arguments, charges and counter charges heated the blood.'[6] But the grant remained owing to the firmness of La Trobe, Victoria's first Governor.

The first St Patrick's was weatherboard and clearly expected to be temporary. Samuel Jackson who had built St Francis's church in the city, was engaged as architect in 1850. His first scheme, similar to St Francis's, was rejected 'as being unworthy of the great patron of Ireland'. His second set of drawings was finally approved - not without misgivings - at a cost of £6000. The foundation stone was laid on 9 April, 1850: 'despite threatening weather the whole of the Catholic population of Melbourne and a goodly proportion of the non-Catholic totalling more than two thousand, attended the scene while banners and flags flew and the Temperance Band played Irish airs.'[7]

The euphoria and the celebrations were sadly premature; poor Samuel Jackson's cathedral was destined hardly to get off the ground. At first, slow progress was blamed on the Gold Rush where the diggings proved more rewarding for the site workers than digging church foundations in the city. Labour shortages suspended building work. But was this merely an excuse for a growing lack of enthusiasm for Jackson's designs? Some accounts suggest so. The building might have been sufficient for a provincial town as Australia's foremost architectural historian J. M. Freeland noted, but was 'entirely inadequate as a cathedral in the most thriving city of Australia.'[8] Freeland's criticism of Jackson's designs and his summary of the first St Patrick's torturous beginnings ends on an optimistic note: 'So was born the ugly duckling which was destined, in a roundabout way, and after a series of vicissitudes, to become the immediate forerunner of the greatest Catholic cathedral in Australia.'[9]

Fortuitously, the architect who was destined to bring this about and justify Freeland's judgement, and realise Goold's dreams, sailed into Melbourne waters at the opportune time. It was only necessary for him to become acquainted with his patron.

Wardell was not the first architect to establish himself in the young colonial city. That honour belongs to another Englishman, Joseph Reed, who arrived in Melbourne six years earlier and designed some of Melbourne's finest buildings including the grand, classical-style Public Library. Reed resented Wardell's arrival and was jealous of his position of influence. He would lead the attack on him in future years and question whether he was the most qualified and most experienced architect in the colony. Standards of practice varied considerably. Many who called themselves architects up until that time were builders by trade who came to the town from other colonies in the 1840s and 50s lured by the demand for housing, churches and public buildings. Even Samuel Jackson who built both St Francis's church and the Scots Church and - as noted above - the aborted St Patrick's cathedral, was originally a builder from Van Diemen's Land. As Freeland notes: 'Shady practices developed and unskilled work such as collapsing walls and splintering roof trusses brought the whole profession of architecture into disrepute.'[10] As a result an attempt was made in May 1851 to form a professional guild setting standards and fees. The first elected president of the short-lived 'Victorian

Architects Association' was Henry Ginn, a practising architect who had come to Melbourne from New South Wales. Early attempts at Gothic design for churches by local builders had been crude and ill-conceived, unsupported by detailed knowledge of the style. The exceptions were those designed by architects such as Pugin and – until he arrived in Australia in 1842 - Edmund Blackett, who despatched designs from England with only a limited knowledge of the sites.

With the arrival of William Wilkinson Wardell, a new era of architectural practice would develop.

When Wardell and his family disembarked, Goold was not in a position to meet the architect because he was away in Europe. The diocese had been left in the care of his Vicar-General, Dr John Fitzpatrick.

Two days before leaving Melbourne, Goold had attended a meeting to raise funds for what the contributors believed at the time was the completion of the existing St Patrick's. A sum exceeding £1000 had been raised. There is some evidence however, that the Bishop had another plan which he was careful not to voice at the time. Did he have Wardell's imminent arrival in mind?

Dean Fitzpatrick wrote to Goold on 15 September exactly two weeks before the *Swiftsure* arrived. It is clear that he, if not Goold himself, was unaware that Wardell was already on his way: 'An architect named Wardell has sent to your Lordship and all priests here testimonials of a very high character. From his letter to me, Mr Wardell appears to be just the man we want ... I hope your Lordship will see Mr Wardell before he leaves England.'[11]

In another letter to Goold, written two weeks after Wardell's arrival in Melbourne, Fitzpatrick informed his Bishop that the architect had not yet called although he had asked him to do so upon his arrival.

Fitzpatrick did not have long to wait. The date of his critical meeting with the architect is not recorded but Fitzpatrick's next letter to his Bishop establishes that they must have met towards the end of October, or early November. After preliminaries, the discussion turned to the need for a new cathedral. The Dean explained the Bishop's views, and how he envisaged a building on a scale and of a splendour that would outclass any rival. He wanted the Cathedral to be a worthy statement of the position of the Catholic Church in the colony; and not least a reminder of the glories of Christian architecture in Europe. Wardell received the impression that the Bishop was not much concerned about raising sufficient funds although the Catholic population was relatively small at the time - around seventy-seven thousand or one quarter of the population of the State as a whole. The Bishop's idealism and liberality were to cause major concerns for his successors but the outcome may be judged to the advantage of succeeding generations. Goold was likely to have been in accord with Wardell's dictum - echoing Ruskin - that 'we build not for this generation only, nor for the next, but for those who will exist in centuries far removed from us.'[12]

An architect of Wardell's experience must have been gratified at the extraordinary professional opportunity that was placed before him. It is given to few architects to design a cathedral - especially one that appeared to have so few financial constraints attached to it. The only condition imposed on him was to incorporate the foundations and as much of the incomplete existing building as possible, a measure that could be interpreted by the population at large as fiscally prudent. In the event very little was retained.

Wardell inspected the site and made a preliminary sketch which evidently impressed the Dean and won his confidence because he wrote again to his Bishop: 'He has made a present of the sketch which I enclose for your Lordship. I think Mr Wardell is competent to carry on any building that is required in Melbourne.'[13]

Goold, who remained in Europe for a further twelve months, endorsed Dean Fitzpatrick's opinion of Wardell and asked him formally to engage the architect. Although a long way from Melbourne, the bishop was certainly in the right place to receive glowing reports of Wardell's work in England.

On 14 December the Dean confirmed that he had engaged Wardell and added in his letter to Goold: 'I think [he is] a first-rate architect.'[14]

And so the work of designing and constructing a monumental new Cathedral for Melbourne commenced in December 1858. Throughout those early years of construction Wardell worked closely with Dean Fitzpatrick who was fired with enthusiasm for the project. Priest and historian Thomas Boland describes Fitzpatrick as an austere man who, as the Archbishop's secretary, did not make himself popular with the clergy, and was a somewhat aloof figure. But Wardell and the Dean worked well together: they became partners and respected each other. Wardell's inspiration changed Fitzpatrick, Boland argues,

'He found an enthusiasm for the ideal, and a passion for the stone itself that gave a human warmth to his piety. Fitzpatrick's single-mindedness induced a glow in him that radiated to others, impelling them to share his dedication ... He followed the shaping and laying of each stone with almost maternal anxiety. No position was too exposed, no height too vertiginous to deter his supervision. Clad in cassock and biretta he scaled the walls, straddled the beams and swayed over the yawning vault. The workmen were amazed and the worshippers were edified.'[15]

Wardell worked rapidly, and produced a plan that was of greater size and splendour than any cathedral attempted elsewhere in the nineteenth century. A scale model of the completed Cathedral, designed by Wardell and executed by a Mr McGowan for the guidance of Dean Fitzpatrick, now resides in a glass case in the nave. St Patrick's with an overall length of 103.6m could claim to be the largest church completed at that time and in that style within the century. Although the plans would undergo modifications and additions as building progressed over a long period, the first sketches, completed within weeks, were generally adhered to.

We will return to St Patrick's in greater detail in later pages as the Cathedral takes shape, but if we follow Wardell's early months in Melbourne chronologically we should now turn to his next great work, his third most significant Gothic Revival building in Australia, St John's College - or to give it its formal name seldom seen but given in official documents, the 'College of St John the Evangelist, University of Sydney'.

CHAPTER 10
'WHAT YOU DO NOW, DO WELL'

The task of designing an immense cathedral without a staff of draftsmen might be considered as much as any one architect could handle, with its myriad points of detail, its specifications, its engineering problems, its quantity surveying and supervisory requirements. But the Cathedral was a long-term project and the commission at 2½ percent on contracts, payable in instalments over a very long period, would have been insufficient income to cover family living expenses. Wardell needed a constant stream of commissions from good contacts to maintain his family at the level to which it was then accustomed.

Although he was, as we have seen, armed with impeccable testimonials from prominent lay and clerical sources there is no evidence that he had at that stage any notion that he would be welcomed in the fledgling Victorian Public Service. We know that Bishop Goold had been advised in advance of his arrival and it is probable that Wardell expected that his main work in the new colony would be much the same as it had been in England - building churches – an expectation that was to receive almost immediate endorsement.

Scarcely had building work on St Patrick's Cathedral commenced than Wardell, who had opened an office in Melbourne at 46 Collins Street West, was seeking more work elsewhere.

There then occurred, probably by chance, a meeting that seemed heaven-sent. Wardell was introduced to Sydney's well-known pioneer Archdeacon, John McEnroe, who was on a visit to Melbourne at the time. From him he learned that a Catholic College was to be built and incorporated within the University of Sydney. A similar denominational college for the Anglicans, St Paul's, had already been established before Wardell arrived in Australia. By Act of Parliament land grants and generous University endowments were available conditionally and even-handedly to other denominations. Archbishop Polding seized the opportunity and was determined that Catholics should not lag far behind.

In his Pastoral Letter of June 21, 1857 he appealed for funds. He argued that the decision to establish the new College named (not without allusive significance) St John the Evangelist, represented 'a vindication of our name as Catholics from the vulgar slander that we fear, or do not love the diffusion of knowledge.' Polding then claimed that the new College would enable Catholic students to 'now begin on equal terms with others, [studying] universal knowledge, for the benefit of that which is the flower and fruit of all science.'[1] The public responded generously to his appeal and a total of £22,805 was raised or pledged.

The immediate concern of the Archbishop and the appointed College Council was to erect on the favourable site a building worthy of Polding's aspirations and a rival to St Paul's in splendour and tradition.

Edmund Blackett, Wardell's eminent Sydney rival in the profession, had already designed the justly famous Great Hall and administration buildings, as well as St Paul's College and St Andrew's Anglican Cathedral. That he would play an important part in the developing drama of the new Catholic College was yet in the future.

Meanwhile Wardell, encouraged by McEnroe, wrote the first of many letters to the Chairman of the College Council, the Very Reverend Dean O'Connell. In the polite terms employed at the time he explained that he felt a certain delicacy in offering his services as an architect, 'and I should hardly have ventured on it but at the Archdeacon's suggestion'.[2] He explains that he is not a stranger to this class of work 'and indeed have had perhaps as much experience and practice in Ecclesiastical buildings as any architects in the old countries'. He concludes that if favoured with the commission to design St John's he will 'do my utmost to acquit myself satisfactorily'[3] and he agrees to come to Sydney to inspect the site.

Then occurred the first of what was to become a litany of delays and misunderstandings in dealing with correspondence. The Building Committee appointed by the College Council did not immediately acknowledge Wardell's letter but waited until after it had reported back to the Council at its meeting on January 4 – eight weeks after Wardell had dated his letter. The Council learned that members of the sub-Committee had met with Archbishop Polding who spoke well of Wardell.

'His Grace thinks the Council would do well to employ that Gentleman as he has come from England with the highest testimonials from Church dignitaries.'[4] Polding as we know was aware of Wardell's work in England and of his association with Pugin.

Meanwhile Wardell in Melbourne was growing impatient for an acknowledgement of his application. To add to his uncertainty, in January he learned from a Mr O'Sullivan visiting from Sydney that the College Council expected the architect in Sydney daily. But he had heard nothing of this. 'This has made me anxious,'[5] Wardell writes in a second letter to the Council, pre-empting by two days the Council's meeting in Sydney that would confirm his engagement. He suggests that perhaps a letter from the Council had gone astray for he was still awaiting their communication.

Such is the nature of committees, responsible and well-meaning though they may be, but meeting irregularly and invariably inexperienced in the business they are responsible for. Inevitably their procedures and delays and misunderstandings will tax the patience of the professional artist and craftsman. Thus the seed was sown in January 1859, dormant and undetected at this early stage, which would sprout later into an unfortunate breakdown in the relationship between the Council and their architect. The great distance which lay between Sydney and Melbourne – seeming greater in those days than today - and the difficulty of communication, would provide fertile ground for the growth of recrimination and dissatisfaction on both sides.

Had the delay in communication led Wardell to give up hope of the St John's commission? Probably; and this may have been a factor in his decision to seek a position in the Public Works Department [PWD]. Although his subsequent appointment as Inspecting Clerk of Works and Chief Architect was Gazetted on March 11 *after* he had received his commission from the College Council, he certainly would have submitted his Public Works Department application and examination designs *before* he knew of his PWD appointment, and before he knew of his commission as College Architect on February 22. Although Wardell as a Public Servant retained his right to private practice, his government work would claim most of his time, (a fact neither anticipated nor disclosed when he first applied to St John's College Council) - a major reason for the eventual breakdown in the relationship between architect and client in Sydney.

Despite the demands of his new career with the Public Service, Wardell was able to write to Dean O'Connell on 9 May confirming that he had sent off the general plan for St John's 'by steam boat', and he hoped that 'they will meet with your approbation.'[6]

Wardell's design called for a very large and imposing college built on three levels in local stone throughout and in the Gothic style, set well back on high ground behind the University's main building. A dominant feature would be the tower over the main entrance which was to rise to a height more that two and a half times the height of the chapel and ridgelines of the main building. The plan form was an H-pattern where the horizontal bar of the H is longer than the perpendiculars. The perpendicular wing on the right of the H - when viewed across the rising grounds from the University - comprised the formal entrance under the tower, the grand staircase leading up from the ground floor to the Chapel, the main hall and refectory. The left-hand wing viewed from the same perspective would comprise mainly student accommodation; and the cross-bar of the H on the ground floor was devoted to administration, reception and professors' rooms. On the first floor were situated the Library and ante room, and on the top floor additional student accommodation. In an early idealised sketch of the College, Wardell shows how he was influenced by Pugin's description of the ideal colleges of Oxford and Cambridge in his *True Principles of Pointed or Christian Architecture*:

'The main feature of these buildings was the chapel: to our Catholic forefathers the celebration of the Divine Office with becoming solemnity and splendour formed a primary consideration ... the place set apart for this holy purpose generally towered over the surrounding buildings. A very characteristic feature of the old collegiate buildings is the position of the chimneys, which are made to project from the front walls of the buildings. The advantages of this arrangement are as follows:

1. All the internal space usually occupied by chimney stacks, and which is very considerable, is gained to the apartment.

2. The stacks thus placed act as buttresses to the walls.

3. The danger of fire consequent upon chimney flues passing through woodwork of the roofs is entirely avoided.

4. A great variety of light and shadow, and a succession of bold features, are gained in the building.'[7]

The exterior of St John's, especially when viewed from Missenden Road, shows how thoroughly Wardell had followed Pugin's principles. St John's, even with later additions and a truncated tower, is uncannily reminiscent of an Oxford college, with its rich ochre-coloured stone catching the sun. The three floors of deep-recessed mullioned windows, and the chimneys which project from the walls and rise to the height of the roof line, all contribute an atmosphere of inner peace, dignity, and ancient scholarship. The tower with its pinnacles and crockets, although well short of the height that was first proposed, manages to rise above the roof line and acts as an imposing landmark in the surrounding urban chaos. Unfortunately from certain angles it looks squat – as indeed it is – and out of proportion to the main building, a poor relation to the Magdalen College tower in Oxford which is illustrated in Pugin's *True Principles* and which clearly inspired Wardell's original plan. Also envisaged by Wardell was a wall enclosing a quadrangle where now stand the open railings bordering Missenden Road. Since this was written a major building program has commenced enclosing the quadrangle and providing additional student accommodation.

The Council Fellows received the plans with qualified approval. They raised several objections and queries which the Chairman of the Building Committee, John Gorman, listed in a letter to Wardell dated June 18. Chief among the concerns was the positioning of the Chapel, 'Considerable objection is taken by the Fellows to having the Chapel on the first floor instead of the ground floor.'[8]

Another concern was the estimated cost of £30,000 and since Wardell had suggested the tower need not be built to its full height until money became available, the Fellows wanted to know what height was included in that estimate. They asked, through Gorman, whether the incomplete tower would have an unsightly effect, and whether the plans could be altered so that the Chapel be sited on the ground floor.

Wardell wrote back on July 6 patiently dealing with each of the objections in turn. This measured courtesy is consistent with what we know of him so far, and although insisting on his principles, he understood that an architect's duty included the requirement to justify his proposals and support them with adequate explanation. He tells the Committee that it is difficult to estimate an accurate cost until detailed specifications are prepared but 'from my experience in such matters and also to the best of my judgement and belief, I am of the opinion that the portion referred to could be erected for £30,000.' He then reminds the Committee that tenders will be called for a 'lump sum' and they would then be in a position to know 'to a sixpence' what they were called on to expend before they entered into a contract.

Then he deals with the charge that the tower, unfinished, may be thought unsightly. He denies this:

'The Committee will, I believe, share my opinion when they recollect the numerous unfinished towers in the old country and how very far from unpleasing they are. I believe the reason for this to be that when the eye perceives that what has been done is right in itself, and it forms a portion only of the work to be completed it is satisfied: whereas a finished construction of a temporary character is always unpleasing for a contrary reason. We have heard many say "What a pity that tower cannot be completed" but I, at least, have never heard it said "how unsightly that unfinished tower is."'[9]

And then he deals with the positioning of the chapel on the first floor. He begins by agreeing that it could be placed on the ground floor but would involve -

'both an additional outlay and to my mind a disadvantageous derangement of the design. Nor do I think that even if these objections were not consequent, no alteration as to position would be desirable. The rule is that the Chapel should be on the principal floor and that there should be no rooms over it.'[10]

Wardell then finishes with a remarkable passage reflecting his deeply-held principles, echoing Ruskin and others, but clearly a personal testament surely deserving of a place in any architectural thesis or reference work:

'You are about to build not for this generation only, nor the next, but for those who will exist in centuries yet far removed from us; and you have with an admirable zeal proposed a work which will vie with the noblest of those edifices that bless and grace the souls of the old countries. But at present your means do not correspond to the full extent of your wishes.

Forgive me, if the interest your kindness has permitted me to take in this good work, urges me to intrude this advice upon you – What you do now do well – even if the funds at your immediate disposal require it to be less in quantity than your generosity intended.'[11]

'What you do now do well' has become a College mantra and is reproduced today in prospectuses and various promotional material. One hopes that the alumni and present-day students know and remember the origin of those words and bless the man who first penned them.

Wardell must have convinced the Council Members for he succeeded in silencing their objections - at least for the time being.

Wardell made a visit to Sydney and established cordial relations with the Council. Work on the College foundations began in mid 1859. But there were storms ahead which would end in a regretful parting, although in fairness to both parties, neither the architect nor the client were to blame. The full story will unfold in later chapters but in the meantime we have to return to Melbourne to follow Wardell's path into the Public Service, his rise to prominence, and the sectarian opposition that he had to contend with.

CHAPTER 11

'A QUARREL OF THE BITTEREST KIND'

The Victorian Government Gazette of Friday, 7 March, 1859 includes the terse statement that William Wilkinson Wardell was appointed Inspecting Clerk of Works and Chief Architect in the Public Works Department. It was signed by the Commissioner of Public Works, George S. W. Horne and authorized by the Governor in Council. Henceforth Wardell would be a tenured Public Servant with his first loyalty to his Government Department.

The appointment was largely dependent upon the outcome of a competition in which candidates for the post had to submit designs for future buildings on the Treasury Reserve. Wardell's design and his application have long since disappeared but a clue remains in the following statement that the surviving Treasury buildings in Melbourne were based on his submission. 'It was stated in the House and generally understood that the Treasury now in progress of erection forms but a portion of the design – it is not intended now to substitute any other style well adapted for the purposes required and which has already been selected.'[1]

Just why Wardell decided to apply for a full-time appointment at a time when he was intent on building up a successful private practice, and had commissions demanding his attention, is not clear. As noted above, he, like other members of his Department, retained the right to private practice provided such work did not interfere with his Departmental duties, but the amount of time he could henceforth spend on private work would be severely limited and was bound by regulation. We can only suppose that Wardell, having explored the possibilities, judged that in spite of his work on St Patrick's Cathedral (a slow and uncertain project), and a hoped-for but uncertain commission for St John's College in Sydney, he could not anticipate enough private work to maintain him and his family in the style he had chosen. He had grown accustomed to gracious living and gentlemanly status.

We have shown that Wardell was ambitious but his ambition was not defined solely by money and status. It was more likely to be a professional ambition – a desire for fulfilment as an artist. Domestic architecture, schools, banks and convents do not generally bring lasting satisfaction to an ambitious architect, nor bring him acclaim and secure him a place in history; but public buildings do. Wardell, still only thirty-five years old, must have realised that to be a pioneer and a powerful influence in the building and appearance of a nascent city – a city to become one of the finest in the New World - was an irresistible professional challenge and one which his experience and talent well fitted him to meet.

Wardell's department in the expanding Victorian public service was, at that time, only three years old and therefore hungry for experienced staff. Its policies and character could be shaped by strong, wise leadership. This Wardell was capable of giving. His proven record both as an engineer and an architect, and his firm leadership qualities were recognised and would lead to rapid preferment and regular steps up the ladder in the ensuing years.

Inevitably this success aroused jealousies and sectarian bitterness ever bubbling below the surface of Victorian life at that time. It didn't help that only six months prior to Wardell's arrival the notorious 'O'Farrell affair' in Sydney had gripped the colonies with fear and panic and was still much alive in the public memory. Rumours of Irish inspired revolution circulated widely and many Victorians anticipated a Fenian invasion. A young Catholic Irishman, Henry James O'Farrell, claiming revolutionary accomplices (later denied) had shot and slightly wounded Queen Victoria's youngest son, Prince Alfred Duke of Edinburgh who was making a visit to Sydney. The Prince escaped serious injury but relations between Irish Catholic and English Protestant did not. Although Wardell had no Irish connections, in the public perception of the affair and its aftermath there was little distinction to be made between being Irish and being Catholic: they were 'all trouble-makers' and suspected of disloyalty to Queen and country. In his authoritative account of the O'Farrell affair, Keith Amos states that the ensuing political furore 'demonstrated with remarkable intensity, the breadth and depth of long-held ultra-loyalist fears and suspicions that radical Irish nationalism constituted the most dangerous subversive influence in Australia.'[2]

The Melbourne *Age* newspaper which acted as the mouthpiece of Protestant resentment displayed a noticeable bias, both in published letters (identities hidden by pseudonyms) and in its policy as revealed in its editorials. Following a correspondent's spiteful attack on Wardell in an earlier issue, an editorial pompously asserted: 'With Romanism at the altar, the Age has never presumed to interfere, but with Romanism on the political platform it has a quarrel of the bitterest kind.'[3] The inference here is clear: 'Romanism on the political platform' covered a broad range of sinister activities and however innocent in intention, Catholic Action was deemed unacceptable by the opposing party. Wardell might have expected that he had left behind the often abrasive relationships between Protestants and Catholics in England following emancipation but such expectation was not to be realised. He found himself confronted by an even more virulent sectarianism in Victoria which would discomfort his progress in the Colony. Four weeks after the publication of the Public Works Department vacancy in the newspapers, and before the announcement of Wardell's appointment was made, a letter appeared in the *Age* accusing the selection committee of having ignored other applicants' submissions and secretly chosen their favoured candidate. Without naming Wardell, yet clearly aimed in his direction, the letter concluded: 'Now Sir, this office [Inspecting Clerk of Works and Chief Architect] is already intended for an individual who *professes* to be an architect, but is not, but having the gift of the gab he has managed to earwig the heads of the Department.[4]

This was followed four weeks later with a specious attack on the integrity of Wardell supposedly written by a disappointed architect admitting he had been a candidate. The letter was signed quaintly, T. Square. The pseudonym and the content suggested a mixture of politically motivated and religious jealousy and hints of newspaper contrivance. The paper's improbable critic claimed that 'The lucky individual who was to enjoy the office as a sinecure, had not only been appointed in England ... but sent out by his Reverend Bishop Goold [sic] for the benefit of his own health and the good of the Church'.[5] The writer then asks: 'Is he not to give a portion of his public time – paid for out of the Treasury – to pull down the Cathedral of St Patrick's built but yesterday on the Eastern Hill and re-build it free of professional costs; in fact this office competition is a mere sham, whereon to hang a thousand a year to pay a Cathedral architect.'[6]

It didn't help Wardell's cause that the leading ministers, Sir John Shanassy and Gavan Duffy, were Irish born Catholics, described later as 'certainly political and social radicals'. But it is unlikely that, in this instance, they interfered with the appointment process or were complicit in the accusations laid against either Wardell or the Department. According to one account there had been seventeen candidates for the post. The successful appointee was chosen after a competitive examination conducted by Major-General Pasley and G. S. W. Horne - the latter a cabinet minister who acted as Chief Commissioner for Public Works under whom the appointee would act as assistant.

The attack on Wardell and criticism of his appointment continued throughout the following months, although it was invariably met with a dignified silence. The *Age* newspaper called for an enquiry into the conduct of the Department of Works and this suggestion was eagerly supported by a correspondent, James Remick Gibson, who accused Wardell of unfair dismissal and was given space to air his grievance:

> 'Having been engaged for a short time in [Wardell's] Department and having been discharged without any sort of warning, without a chance of being heard in my own defence, and without any fault, as far as I know, except, perhaps that I was too inquisitive, I hope that this enquiry will take place and that I shall be called upon to give evidence. I can a tale unfold that would open the eyes of the taxpayers of the country. The whole Department is rotten to the core.'[7]

This level of attack on Wardell in the *Age* would continue throughout his Public Service career and would lead eventually to a Parliamentary Enquiry and a Royal Commission. In the meantime members of the Government preferred to read the conservative *Argus* which was generally contemptuous of its rival.

The sniping had no effect in preventing Wardell's promotions and increasing his responsibilities which continued throughout his first two years of office. Within one year of his arrival in the colony he had become a powerful voice in the Victorian Public Service, holding down a string of appointments to various boards and enquiries. On 1 June, 1860, six months after his initial employment as Inspecting Clerk of Public

Works, he was given the title Acting Inspector of Public Works and Chief Architect. After another six months he was promoted Inspector General of Public Works and Chief Architect. He was then effectively the head of a large and influential Department responsible to a Government Minister, and a powerful voice in the shaping of the city. His many extra-curricula appointments around this time included Trusteeship of the projected Zoological Gardens; Membership of the Board of Examiners of Land and Works; Member of the Commission to inquire into Fine Arts in Victoria, and a Member of the Commission to inquire into the Yarra Bend Lunatic Asylum.

All this preferment of a known Catholic served to exacerbate sectarian bitterness. Under the State Constitution - unlike in the Mother Country - there was no special status accorded the Protestant Church. This at any rate may have been the founding fathers' ideal. In practice members of the Protestant community continued to maintain their right to the dominance that they had enjoyed in England. And the *Age* was the most shrill voice in support of this view:

> 'No careful observer can doubt that the one great feeling that permeates the Protestants in Victoria at this moment is the urgent necessity for co-operation amongst themselves to counteract the machinations of the Pope's servants who now hold the reigns of power, and whose evident design has been, is, and ever will be, to secure Catholic domination. Once let the R.C.s number more than one third of the entire population and their perfect organisation and unity in slavishly obeying the bidding of their priests, will give them a virtual majority; and depend on it, Sir, not long after that sad event, Pope John with his Tipperary boys may be employed in razing to the ground some of our fine Protestant churches, or assisting at an auto-da-fe of Bibles in the public streets. In ordinary times bigotry and suspicion are to be dreaded and shunned; but there are exceptions to every rule and this is one of them. Every Government office is literally swarming with Catholics – chiefly Irish Catholics. There is a distinction in favour of them.'[8]

Three days later a letter appeared, signed simply, 'An Englishman' in which the earlier letter was described as an eloquent and stirring appeal which 'will prove in rousing the Protestant co-religionists to a sense of the present sinister aspect of affairs in the Government of this country.'[9] It appears not to have been the custom for the *Age* of those days to give space to letters expressing contrary views.

If Wardell read the attacks on him in the *Age* there is no evidence that he was greatly disturbed by them. He had powerful supporters in the Government and at that time there seemed no reason to believe that his position was other than secure. He began to lead a busy social, as well as professional life.

The letters describing his impressions that Wardell sent back to his old associates in England shortly after his arrival are, regrettably, untraceable now. But we get a glimpse of him in the replies from friends that still survive. His artist friend David Roberts acknowledged Wardell's letter written just two months after arrival, saying he was -

> 'most happy to learn from our friend Stanfield that you have fairly settled down and [are] enjoying your new adopted country. I hope to hear you may have received health and energy to carry out those talents you carry with you.'[10]

Wardell had asked Roberts for advice in obtaining casts for a proposed School of Art and National Gallery. Evidently Wardell had identified the need for such a school to train local artists and architects rather than relying on the irregular and sometimes dubious candidates from overseas. Wardell showed that he subscribed fully to the art training orthodoxy of the period. He enquired of Roberts and Stanfield how he could obtain the best casts and antiques as models for students – a priority for sound instruction and practise. Casts of the Elgin Marbles were discussed as a possibility in Robert's letter. Roberts suggested Stanfield could obtain these for him.

For how long Wardell pursued his ambition to establish an art school is not clear; there are extant letters from his English friends on the subject, but the rapid increase in his professional commitments precluded continuing work in this area and the subject soon disappeared from correspondence. Later, Wardell was appointed to the Board under the chairmanship of Sir George Verdon whose purpose was to establish what became known as the National Gallery of Victoria.

Among the first churches that Wardell designed in his new country were St John's Catholic Church, Heidelberg; the Anglican parish church of St John's, Toorak; and the first St Mary's, East St Kilda - the latter becoming his parish church which he substantially enlarged when he took up residence in a spacious family home, 'Ardoch', 226 Dandenong Road. This elegant Victorian-style house is still there, now divided into two flats and incorporated into a small estate of extensive private residences.

The Wardell parish churches of Victoria - near fourteen in total excluding the Cathedral - are far simpler in construction, smaller, and may appear less interesting than the parish churches that he built in England. The reasons are not hard to appreciate. There was far less money available for church building at that time, the population was not only much smaller but scattered and consisted in the main of Irish manual workers. There were no rich patrons in the wings, and building skills and craftsmanship were harder to come by. Wardell therefore adapted his designs with these constraints in mind. He turned to Early English principles rather than Decorated Gothic favoured elsewhere. The church interiors are simple in form, plainer and lacking 'Puginian' decoration; nor are they innovative in any degree. But are they less interesting? No! Even the simplest of them has an individuality and beauty of proportion and recognisable Wardellian features. Each church is structurally sound, and aesthetically pleasing because, as Ursula De Jong has so accurately observed,

'Wardell has a remarkable ability to handle his materials of timber and natural stone with sensitivity and mastery. And believing in the inherent qualities of the materials themselves, ornament flowed only as an outgrowth of those qualities.'[11]

The first St Mary's in East St Kilda, Wardell's first parish church commission, arose from his newly established friendship with Dean Fitzpatrick. So impressed was Fitzpatrick with Wardell's plans for the Cathedral that he strongly recommended Father Niall and the parishioners of East St Kilda to engage that architect for their new church. Commenting on a plan drawn by a builder named Scanlon, the Dean wrote to his Bishop, 'I gave them a good scolding and told them it would be better to burn their money than expend it on such a horrid building, I advised Fr. Niall to engage Mr Wardell to prepare a suitable design.'[12]

Wardell's first design for the church is dated December 1858, only two months after his arrival in Melbourne. Compared with the later church (1869) the earlier version was smaller; the bellcote and west end is there, but the nave has only five bays, has no porch or side chapels, and only a simple four-square chancel with a lower roof line, reminiscent of St Mary's Axminster. The later design involved considerable rebuilding: the nave was lengthened by three bays, various features were added – an apse, apsidal chapels and a choir chamber. The roof line of the chancel was extended to equal the height with the nave. The interior of this, the second St Mary's is the most satisfying and aesthetically pleasing of all Wardell's smaller parish churches in Australia. The reasons are not hard to find: St Mary's was Wardell's own parish church and he lived only a short walk away along Dandenong Road. Thus he was on hand to supervise its construction - not always possible with his other churches because of Public Service rules determining the amount of private work staff could undertake. Another important consideration was the support and intimate friendship of his new parish priest, Father James Corbett, a cultured man, a liturgical expert who was devoted to Gothic Revival architecture.

The aspect when entering the church is little short of sublime; the proportions, the harmony between the vertical and horizontal lines are so perfectly realised. The eye is immediately focussed on the pillars and the graceful arches of the bays which lead upwards to the timber vault and then along the horizontal towards the chancel. Spanning the division between nave and chancel - there is no chancel arch - is a dominant feature: the rood set on its curved beam reminding us that Wardell, like Pugin, favoured the traditional rood screen - although seen here in a much simplified form. Across the wooden beam are painted the dramatic last words of Christ on the Cross, *consummatum est*. Rising up from the centre is the huge crucifix. The chancel is lit by three stained glass lancets and two of plain glass either side, set around the curved apse. The nave is lit by aisle lancets and quatrefoil clerestory windows above. Apart from the screen and sanctuary furnishings the rest of the church is minimally decorated. Its sublimity lies more in the harmony and careful proportions of the different architectural features. The result is peaceful, prayerful, and satisfying.

Visitors should not leave St Mary's without first paying homage to the architect whose likeness, based on a photograph, may be found depicted in a stained glass lancet at the west end of the south aisle. The figure represents St Thomas the Apostle who is believed by tradition to have been an architect who brought the Gospel to distant lands and built churches. He carries a t-square and staff and is given a halo, but in this depiction his head is painted to look remarkably like Wardell. Wardell the historicist would have been well aware of the historical custom of using living people for models in mediaeval iconography, but perhaps might have demurred at standing in for St Thomas. Teresa Wardell and friends sponsored the window as a memorial to her grandfather when the church was restored in 1984.

St Mary's, the most personal of Wardell's parish churches, displays unerring taste and deep spiritual conviction. The architect seems to be saying to succeeding generations, 'this is what a modest parish church should be like'.

CHAPTER 12

A CONFLICT OF INTERESTS

Architects can and often do design buildings which are then built far from their home practice; some have only a scanty knowledge of the site and, after despatching the drawings, have little further input. In these cases they then rely on a local architect to supervise the construction based on their original designs. This was widely the case in nineteenth century Australia when access to competent architects was difficult, or when a notable overseas architect like Pugin or Butterfield was preferred as a matter of prestige. But with important, large public buildings the client was justified in expecting that the architect would be on hand, or at least readily available for consultation throughout the building process. Wardell would have endorsed this view as is evident from his close involvement in the building process of his two Cathedrals and other buildings. Only in one of his major buildings, St Mary's Cathedral, Hobart, was he not on hand to supervise and that, as we shall see, had disastrous results.

The fact that he was not able regularly to visit the site of St John's College within the University of Sydney, was also a cause of misunderstandings and mounting conflict between architect and Council towards the end of 1869. Throughout this early stage the Council was impatient and pressed for the building work to begin. Although the delays were not expressly Wardell's fault, the difficulty of communicating with him, the irregularity of Council Meetings, and the Council's lack of sympathy with Wardell's reluctance to rush the creative process, all contributed to a worsening situation.

Our sympathies alternate between architect and client; the client impatient at the delay demanded detailed drawings so that tenders for carrying out the works could be called. The architect protested that their time scale was impractical. Wardell wrote:

'I am sure that two months of uninterrupted labour is the very minimum of time in which [the plans] could be completed and it would be a matter of very great regret to me if I were limited to that. The plans require work that requires considerable thought and reflection and I am sure you will agree that it would be very undesirable to hurry over. I venture therefore to ask you for three months.'[1]

Wardell's professional life at that time must have been feverishly active when we consider that his dominant responsibility was to the Public Works Department and only in his spare time could he attend to private work. This was to be spelt out unequivocally in a memorandum from J. S. Johnston, the Commissioner of Public Works when the professional magazine, *The Australian Builder*, raised objections to Government officers of the Department accepting private work. Johnston wrote:

'Private practice is distinctly forbidden during the hours of business ... the hours fixed for business of the Department are to be strictly and faithfully conserved for that alone. When [the officer's] services are not required, the hours of leisure may be used for his own amusement or profit provided no supervision of any work he designed be undertaken.'[2]

This must have proved a severe limitation to Wardell's extra-curricula activities. He was unable to absent himself from his Department to make trips to Sydney when the Council felt they needed to consult him. But there was another reason for his difficulties which was equally important to Wardell. He was a conscientious Catholic who followed strictly to the letter the Church commandment to avoid servile work – that is, manual work and work for profit – on Sundays. Yet the only day that he found himself free to work on his church designs was that day. Those ancient prohibitions are largely forgotten or disregarded today, but for Wardell and the majority of Catholics of that time - and indeed for most other Christian Churches - Sunday was sacred and a day of rest. To solve his dilemma family legend tells how he sought and received from his Bishop a special dispensation to allow him to work, on condition that Sunday work was limited to commissions for the Church.[3]

The St John's College Council was not as understanding. The ceremonial laying of the foundation stone of St John's was arranged for December 27 and it was expected that the architect would attend. Wardell was given leave from his Department, largely, one suspects, because it was

Christmas time. He wrote to accept his invitation and adds that 'it would be the greatest gratification to me if I could, in any way, be useful either in the ceremony, or in preparation for it.'[4] He then adds that he will be in Sydney a few days prior for that purpose.

This seemingly innocent last sentence provoked an angry telegram response from John V. Gorman, Secretary of the Building Committee. 'Council considers your letter of 23rd [November] unsatisfactory. You must come at once. If your engagement with the Victorian Government prevents your proper attention to your work, it will be a breach of engagement with us.'[5] Wardell, normally so proud and confident, is obviously greatly hurt by Gorman's response. He wrote back to explain that he could not leave Melbourne before December 16 but he was confident he would be able to make every arrangement in ample time for the ceremony on the 27th. He added in his telegram 'If the Council wishes to dispense with my services I am quite ready to give place to anyone they prefer, on the other hand am equally willing to fulfil my engagement with them to my best ability.'[6]

This exchange of telegrams marks a turning point in relations between architect and Council from which they were never able to recover. It may be deduced from subsequent correspondence that the Council members - men of high standing drawn from the law, and from commanding positions in the Church and academia - viewed Wardell very much as an artisan or a superior tradesman; theirs to command. But they were dealing with a proud man, an artist, not a servant of anyone, and one who was very conscious of his worth, his long experience, and his status in the profession. A clash was inevitable.

Differences were not likely to have improved when it was learned that Wardell could not get a boat leaving Melbourne until Monday evening, December 19, which would not reach Sydney until Thursday morning December 22, five days before the ceremony was to take place. Even allowing for this delay Wardell was able to complete arrangements for the elaborate ceremony performed by Archbishop Polding. The Archbishop had insisted that an outline of the building and foundations should be dug in preparation. Work on the College had to be seen to have started. The architect was welcomed, lengthy discussions took place with the building committee, and peace reigned - for the present anyway.

While in Sydney Wardell arranged with architects Weaver and Kemp of Pitt Street to inspect and superintend the works on his behalf and visit them once a week. 'I need not say' wrote Wardell, 'that in addition to their inspection I will visit the works as often as I properly can.'[7] He agreed to pay Weaver and Kemp out of his legitimate architect's fees which he reminds the Council are due - £300 as a first instalment and then every three month as the works progressed.

A month later he complains to the Council that he still awaits payment: 'I may say that I have already paid out of pocket to draughtsmen and their expenses over £120 on account of this work...this is nothing more than what is fair, just and usual.'[8] By 21 March, 1860 he thanks the Council for the payment of £300 but the payment of future amounts continues to be the subject of delay and increasing controversy.

Further criticism is levelled at him for what the Council judges to be an unnecessary delay in receiving the detailed plans for building work to begin. He responds by accusing them of being 'a little unreasonable in being 'greatly dissatisfied' at the plans not being sent. 'I explained in my letter of October 3 that I should not like to promise them under three months.'[9] Whenever he writes to the Council during these months, on whatever point of detail, he invariably ends by 'begging' to remind them about the amounts still owing to him for work completed. He resents having to argue for fees which are prescribed by his profession. Council procedures for issuing payments were cumbersome and authorisation had to await infrequent Council Meetings. The question of non or delayed payment of fees was to became a major point of disagreement between them. In one letter Wardell states he is willing to modify the terms of his original arrangement and 'cut the rate from 4% to 3% on all payments made for the works.' However this olive branch was not strong enough to retain their confidence. Wardell was later enraged at what he judged the final insult to his integrity when he learned that Gorman had written to architects, Blackett, J. F. Hill, and Weaver and Kemp, to check the accepted professional rates for architects - in effect to check on Wardell's estimates. To add fuel to the already smouldering fire the Council forwarded to Wardell the architects' report. In Wardell's angry letter of May 16 he accuses the Building Committee of giving the impression 'that they had met a Sharper who had deceived them, and was trying to do so again.'[10] He reminds them,

'My own standing in the social scale, as well as in the profession is equal to that of any of these [the architects consulted], and my acquaintance with the practice of Architecture as a profession, is probably as much as all four of them.'[11] He concludes his letter by repeating his offer of resignation but adds that the work itself is ' no common interest to me and I should be sorry to be disconnected with it so long, at least, as I can continue it without a sacrifice of my self-respect.'[12]

A month later Wardell has not received any acknowledgement of his offer to resign, or the apology he had asked for. 'I have again read the report signed J. V. Gorman' he wrote,

> 'and the perusal makes me the more urgent in my request that the Council will accept the resignation I have offered, or make me a distinct disclaimer of the imputations that document so offensively conveys ... I presume pending their answer, I have their permission to consider my official relations with them suspended.'[13]

On receiving this letter, the Council Minutes noted that any final settlement of the matter be postponed 'until Mr Wardell would be able to favour the Council with a personal visit.'[14]

And so the impasse continued. Looking back on the correspondence from this distance it is probably fair to say that neither the members of the Council nor Wardell really wanted the break-up to occur but Wardell, like an ancient duellist, believed he was owed an apology - his good name and standing in the profession had been impugned. But the Council would not agree that an apology was justified. The severance which had been long in coming, was, in the end, inevitable. A Sub-Committee had been appointed by the Council to review their relations with Mr Wardell and their conclusions were tabled at a meeting on 18 July. The report concluded that it was 'highly inconvenient if not impossible that such work as the Council is engaged on should be carried out in the absence [of the architect].'[15] The Committee then observed that the difficulties had been caused mainly by 'Mr Wardell's acceptance of a Government appointment in Victoria [after] he entered into the agreement with the Council.'[16]

There were other exacerbating factors. As misfortune so often steps in at the worst times - one woe treading on another's heel - the detailed plans and specifications which Wardell completed successfully were detained by the Post Office in Melbourne because of a foolish blunder on the part of a junior who attached insufficient postage on them. By chance the error was discovered but the consequent delay added further to the already hostile atmosphere. And again, Wardell had engaged a Clerk of the Works, John Denny, whom he had known and trusted in England and had then encouraged to migrate to Sydney to work on St John's College. It was neither the Clerk's fault, nor Wardell's, that the ship bringing him was much delayed and the date of arrival was several times postponed – again, a provocation to impatience by the Council.

Finally, on 28 July, 1860, eighteen months after his engagement, the Council met and accepted Wardell's resignation. However, a long time elapsed before alternative arrangements could be made, and before the resignation was reported publicly - if the cryptic confirmation of it a month later in the Irish Catholic *Freeman's Journal* is any guide. According to the *Journal*,

> 'The original plans of Mr Wardell, for the erection of St. John's, are at present under the consideration of Mr Blackett, and as soon as some modifications of the plan can be adjusted, the contracts already received shall be, if compatible with the proposed changes, received: if not fresh tenders are to be invited.'[17]

In spite of the resignation, while the building of the foundations continued, the discordant correspondence between Wardell and the Council dragged on. In acknowledging the sub-committee's report which the Council had forwarded to him Wardell protested at the implication that he had broken the original agreement - 'unfounded in fact.' He also protested at the recommendation to withhold a portion of the fees 'which I have earned and have a legal right to receive'.[18] He denied that any inconvenience or extra expense, as was claimed to have occurred, had been occasioned by any act on his part.

Not until twelve months later, after persistent demands from Wardell, did the Council finally pay the money to which the architect believed he was entitled. In one desperate letter he wrote: 'Whatever misunderstandings may have existed I cannot but believe that the Council will desire to do me justice'.[19] He then points to the fact that the works are currently proceeding from his plans and specifications which he had supplied – 'the best proof that I have earned the money'. He continues, 'I feel sure that a body of Catholic gentlemen will not wish to withhold its wages from the workman.'[20]

Exactly one year after his resignation was accepted the Council remitted what it owed. Wardell acknowledged receipt of £450 on account, 'which together with £300 paid and acknowledged before makes a total of £750 received by me.' Even here he points out that he has been underpaid by £143.17.0 on the full contract price of £35,754. As we have noted elsewhere Wardell was meticulous in demanding his rightful fees throughout his professional career but it must have been distasteful for him so persistently to argue for every penny. Whether he ever received this final amount claimed is not clear from the correspondence.

Today, one hundred and fifty years later, the long, painful squabble between the Council and its architect has largely been forgotten. Although upon Wardell's departure Edmund Blackett was appointed to oversee the work, he was permitted to make only a few minor modifications to the original. With pride and celebration, Wardell is acknowledged as the true, foundation architect of St John's, the most beautiful and aesthetically satisfying among the admired colleges of the Sydney University. Even the original Council members must have realised, as the structure progressed, that they had an architect and a building of very exceptional quality for they insisted that Wardell's plans be adhered to. They would not, for example, allow Blackett to substitute wood for the more expensive, grand stone-carved staircase, or make other cost-cutting alterations. Wardell's sound advice, 'what you do now do well' was well remembered by the Council and seemed to guide all their subsequent deliberations.

CHAPTER 13
'GRAND IN ITS DESIGN'

Within a few years of his arrival Wardell became one of the 'Clubbable Gentlemen' of Melbourne. There was nothing unusual about this. Most men of the mid 19th century successful in the professions, in politics, science and the arts, belonged to clubs; it was the age of the great clubs in London, and several of these were imitated in Melbourne at different times – the Savage, the Athenaeum, the Union and the Melbourne Club among others.

The first club that Wardell joined was probably the Yorick, so named because of a skull presented by its first Secretary, Marcus Clarke. The Yorick was predominantly literary in character, attracting journalists and writers of a Bohemian leaning. William Wardell is listed among the foundation members in 1868 who each paid the two guineas membership fee, and two guineas annual subscription. The Yorick is chiefly remembered today for the pranks and practical jokes perpetrated by a clique of journalist members addicted to late night carousals - led by Marcus Clarke. On one notorious occasion a rabble staggered up and down Collins Street in the early morning hours interchanging brass plates so that, later in the day, doctors' residences were confused with 'Ladies' Seminaries' and Solicitors with Dentists. It is doubtful if Wardell would have approved of such behaviour which injured the reputation of the Club and was reported by the self-same journalists in the newspapers.

> 'The majority of the Yorick members were staid and orderly men who would not have sanctioned practical joking had they been aware that the Club was made the centre for the hatching of little plots against the public peace.'[1]

Several Yorick members, including Wardell, graduated to the Melbourne Club when invited to join – membership was, and still is, by invitation only. Wardell was admitted in February 1868 proposed by Claude James Faerie, a Melbourne Barrister, and seconded by F. C. Standish, a colourful, elegant epicure of independent mind. He is remembered today as the somewhat ineffectual Chief Commissioner of Police at the time of hunting the Kelly Gang and, less respectably, of being nearly thrown out of the first floor window of the Club by a member whom he had insulted. He spent his last years in residence at the Club and died there in 1883. Wardell would have needed these men as friends to have been proposed and seconded by them. He remained a member for the rest of his life, continuing as a country Member after he moved to Sydney.

The Melbourne Club moved into its neo-classical-style, discreet premises in Collins Street in 1869. There is some evidence that Wardell himself had a hand in the design of what is now regarded as one of the handsomest Victorian buildings in Collins Street.[2] Certainly the exquisite Minton tile flooring in the lobbies must have been chosen as a result of Wardell's friendship with Herbert Minton. Wardell's membership of the exclusive Melbourne Club and his regular visits there would later be singled out in *The Age* as a criticism of his life-style.

Thus within a comparatively brief time after arriving in Melbourne, Wardell succeeded in adapting to his new circumstances and had embraced life in his new country. As the eminent painter, David Roberts, aptly surmised in his letter, he had 'fairly settled down' and was 'enjoying your new adopted country.'[3] Professionally Wardell was to become an influential Public Servant in a position to affect the appearance and the quality of Melbourne public buildings. In private practice he was the creator of a monumental cathedral variously described as 'of global significance',[4] and, 'a Gothic vision probably unsurpassed by any other building in the four hundred years since the great Gothic era.'[5]

Socially, in spite of attacks in *The Age*, he was accepted in Melbourne society and was a prominent figure making many influential friends. His family life and his deep religious Faith sustained him – his prayer books with prayers he wrote himself inserted between the leaves, his bible and his rosary beads, are now on public display with other memorabilia in the Wardell Room of the Melbourne Archdiocesan Historical Commission museum.

Four Australian-born children were added to the family. His fourth daughter, Agnes Mary was born in 1860, Lawrence George was born a year later, and Constance in 1864, and in the following year, his youngest son, Herbert Edward. In total, Lucy Anne bore him eleven children but early deaths, common at that time, reduced the survivors to seven. Michael Thomas, as already noted, died aged five in England; Agnes Mary aged sixteen died in Melbourne in 1876. Two others, Francis Wilkinson, and Lawrence George died in their early twenties. In the Victorian Age child death was accepted stoically but we get a hint of the Wardells' sorrow at the death of young Agnes from a letter of condolence written by the artist John Rogers Herbert. He addresses Wardell as:

'My dear old friend, let the first sentence to you convey the real sorrow we felt in the blow which fell upon you and your dear wife in the loss of your daughter ... nothing really dies except into life. I have had two grown-up treasures swept away and I lean on the truth I state for some solace and find it.'[6]

Herbert's concluding sentence reminds us how popular and well-remembered Wardell must have been in England, and remained so many years after he had left. 'I often wish' he wrote, 'that something could bring you back among us. Things have changed, but you have many friends who would be glad to have you all back again.'[7]

But Wardell had put his hand to the plough, and was not looking back. He had much work to do in Australia, dividing his time unequally between his Government office, his registered premises in the city, and his home in East St Kilda. His grand mahogany, walnut and oak desk which he designed to his own requirements, bears testimony to the many hours he spent in the evenings and weekends working on his church plans in his study at home. The unusual design incorporates pedestals with drawers set in both sides, front and back, where drawing instruments were kept. There is also a secret drawer in which Lucy Wardell is said to have hidden her jewels when going away on holiday.[8] The large desk-top can be lifted and tilted like a draughtsman's easel, its ink stains speaking of its constant use. The priceless heirloom was handed down from father to eldest son Edward, Deputy Master of the Royal Mint in Melbourne. It now rests in the Diocesan museum.

Among Wardell's churches of those early years of the 1860s - doubtless designed at this same desk - were St John's Heidelberg; SS Peter and Paul at Ashby, now known as West Geelong; and his second cathedral, the ill-fated St Mary's in Hobart commissioned by Bishop Willson whom Wardell had known in England.

The designs for St Mary's in Hobart date from 1860 and the ground plan – a great tower over the crossing, and two identical west end towers and steeples - pre-figure St Mary's in Sydney but on a smaller scale. When the incomplete building was partially opened in 1866 it was much admired. The euphoria, however, was short-lived. The foundations were faulty, huge cracks in the walls began to appear, and pieces of masonry collapsed so that the building had to be closed in 1875. Unknown to Wardell, construction work had been sub-contracted to an incompetent jobbing builder by the name of Callam and poorly supervised. Eventually the tower collapsed. Although at the time Wardell was not informed of local building arrangements, and legally could not be held responsible, he feared that his professional reputation might be damaged as a result of the scandal. Some years later we find his explanation of the sorry affair in a Sydney paper, *The Australian Sketcher* of June 1880. Wardell complains to the editor that a report in an earlier edition 'could mislead you and the public, and convey so very unjust a reflection on me.'[9] He then states the facts:

'I made, some eighteen or nineteen years ago, designs for the church; but as I had not seen the site, and therefore knew nothing of the quality of the ground for foundations, I stipulated that I would not incur any responsibility for them and limited my plans to works above ground only. All that was necessary to be done underground was to be designed and carried out under the direction of a local architect. What precautions were taken I do not know, for I have never been able to visit Tasmania. The description in your issue implies that the failure of the lantern was owing to defective work and materials above ground. I think, as my name is introduced as being the architect of the original building, it should in fairness, have been stated that I had no charge or responsibility whatever for the works; that they were carried out under the control and direction of a local architect, and were paid for on his certificate and that I have never seen them.'[10]

Today, with the help of a $1.6 million Government grant, Hobart's St Mary's has been restored and retains elements based on Wardell's original concept.

As important as all Wardell's commissions were to him, the one that came to dominate his private practice and earn him fame was Melbourne's St Patrick's Cathedral. Wardell worked on his drawings and specifications with extraordinary speed, to the point where the foundation stone was laid with minimal ceremony on 8 December, 1858, not much more than a month after receiving his commission.

A cathedral architect's creation, much like a composer's music, begins with marks on a paper. The architect is concerned with hundreds of detailed drawings relating to the various parts of the building and its smallest features. All the while, the architect reads his marks on paper and sees in his mind the intended, finished building - as similarly the composer reads the notes and hears his music. To the uninstructed observer, architectural drawings may be confusing, even tedious, but the architect will see them three dimensionally.

Wardell's instruction from his Bishop to retain as much of the old building as possible was honoured more in the breach than in the observance. His early attempt to utilize Jackson's columns and groined aisle vaults was soon abandoned. A note on one of his plans, dated November 1858, ordered that the old work 'must be removed as soon as possible.'[11] In Freeland's words, 'the second St Patrick's Church in Melbourne vanished from the passing scene.'[12] This did not happen without controversy and accusations of extravagance and wasted effort, but the decision to start afresh has proved in retrospect to be the correct one. Quoting Freeland again, the Jackson church was 'an ugly duckling...a conglomeration of Gothic styles that did nothing to enhance the [builder's] reputation'[13].

Throughout the early years of the 1860s decade, the Cathedral took shape in its dominant position on the Eastern Hill. The soaring scaffolding encasing the giant West Front over-shadowed the few scattered buildings in the neighbourhood - an area of unsealed roads resembling, at that time, a virtual waste-land. In 1862 the facade with its six-light mullion window – as yet without glass - and the north walls rose up to dominate the landscape giving momentum to the appeal for funds. Donors could appreciate the grandeur of what they were being asked to pay for. In 1866 the walls of the nave were up to the height of the roof trusses. The north-west tower, also encased in scaffolding, was rising above the west gable.[13] By 1868, ten years after Wardell's arrival in Melbourne, most of the nave and the aisles were in place and ready for use. The building thus far was proudly put on show when the Bishops of the Australian Province assembled there in April 1868 for their Synod. This important ceremonial gathering could hardly have taken place in the foremost diocese, Sydney, which was still without a cathedral after the disastrous fire of 1865. Although work had started on the new St Mary's it was in no fit state to host a Synod. And so when the Bishops and their retinue progressed through the west portal of St Patrick's and into the unfinished yet soaring nave, Polding must have been envious of Melbourne's success. In Sydney, little beyond the foundations of a cathedral was in evidence. His efforts to raise money for new building contracts were less successful and building ceased. Diocesan finances under Polding were in a disorganised state.

Wardell continued to work on St Patrick's for the rest of his life, even after he had moved to Sydney. He continued to be involved by correspondence, directing instructions to contractors and the clergy, sending modifications, designing furnishings, and answering queries. For this purpose he would journey each year to Melbourne, reside at the Melbourne Club, and check his work. Only five years before his death he was still exchanging letters with Archbishop Carr, successor to Goold. In one letter he expressed his insistence that the incomplete Tower should be immediately covered to protect the internal walls and transepts against storm water damage. He assured the Archbishop that the cost of this work could be kept to a reasonable £268.

But by that time Archbishop Carr was finding it increasingly difficult to raise money for continuing building work and for Cathedral furnishings. He was being pressed on all sides and traduced by those who believed funds were being misappropriated. But Carr himself gave large sums to the Cathedral and he set up a Memorial Building Fund when Dean Fitzpatrick died in January 1890. On the matter of roofing the tower Carr replied to Wardell:

> 'If I had the money for the purpose nothing would please me better than to have it [the work] done. I have made a resolution extorted from me by dire necessity that I shall not incur any further expenditure in connection with the Cathedral, beyond what is absolutely necessary until I see my way to pay my present oppressive liabilities.'[15]

Wardell, on receiving this, was not the man to accept the Archbishop's argument meekly. As we have seen, on professional matters especially, he was high-principled and stubborn. His reply despatched within the week was a model of polite persuasion: 'I never commenced a letter to your Grace more unwillingly than this, because I know how sensitive and anxious you are on the subject of expenditure and this is to urge you to reconsider the decision you gave me the other day – that nothing more was to be done towards roofing the tower'.[16] Wardell succeeded – as he did with most of his clients - in changing Carr's mind on the matter and the roof was added.

Another instance of Wardell's skill at winning over the Archbishop was when the builder, Wright, had persuaded Carr that money could be saved by using cheaper stone for the higher window facings. Wardell would not agree to such cost-cutting, proving that the amount saved would hardly amount to £6. In his letter to the Archbishop Wardell quoted a verse by Pugin:

Some folks they build a front as great
As old Westminster Abbey,
And then they try the Lord to chate
And build the back part shabby.[17]

Carr recognised the humour and the wisdom of refusing to compromise. Succeeding generations are the beneficiaries of that wisdom today.

Externally, and at first glance, St Patrick's may not appear particularly impressive. The undressed bluestone (Basalt) gives it a stern, unsmiling aspect. It presents a somewhat dumpy appearance when compared to St Mary's in Sydney. This is partly due to the transepts having externally buttressed aisles, unlike St Mary's, which add to their width and tend to detract from the overall lateral dimension of the building. Another reason may be the extended height of the three spires, out of proportion to the rest of the building. They were completed only in 1939 and were clearly unintended by Wardell; they lead the eye ever upwards - the lateral line, in effect, minimised in favour of the vertical which may account for the Cathedral's slight dumpy appearance. But on closer inspection and from whatever position the Cathedral is viewed externally, the visual delight is in the complex massing of forms and the variety of vistas – the chevet chapels with their pinnacles and peaked roofs, the massive buttresses and the transepts, the massive central tower with the tallest of the spires rising 344 feet from ground level, the two sister spires of the west towers, all drawing the eye upward until the view tapers away at the pinnacles and the huge cross on the central spire is minified in the clouds. Wardell might have argued that the reinforced tower and higher steeples dating from 1938 spoiled the symmetry of his design and made the cathedral look top heavy.

However inspiring the exterior may be, the real beauty of St Patrick's is revealed when the visitor steps inside the West door. Like opening a large book with an attractive cover, the true delight is in the contents, inside. The vista when viewed on entry by the side portal in the West Front is immediately breath-taking. The great length extends three hundred and forty feet from the entrance towards the east end where the unusual apse comprising seven small chapels in the French Gothic style, seems far in the distance. The great height of the nave, at ninety-five feet higher than Durham and Gloucester Cathedrals, is crowned with a timber vault, angels on the hammerbeams, and a barely visible carved frieze above the clerestory and the width of the nave, greater than Canterbury, Salisbury and Norwich, combines to give an impression both majestic and sublime. The historian and architect, Robin Boyd wrote in 1962,

'Viewed in the golden light of amber glass windows, the Cathedral's interior presents a Gothic vision probably unsurpassed by any other building in the four hundred years since the great Gothic era.'[18]

Before moving up to the crossing, turn around, look up and admire the great west window – the Ascension Window - above the portal. The spectacle of intense, translucent colour, predominantly reds and blues and the systematic curvilinear patterning, is breathtaking. If there should remain any doubt about Wardell's worth as an artist, his design for this masterpiece (not forgetting similarly the north window in St Mary's both windows manufactured by Hardman's of Birmingham), should prove his genius. Again, we see this touch of genius in his decision to design the transepts with aisles so as to have the advantage of additional space that is thus created. The Cathedral is no longer long, dark and narrow (as Butterfield's St Paul's is in the same city), but vast, open, and flooded with sunlight. As in all massive buildings and especially cathedrals, the visitor is hushed by the quiet stillness and the harmony of the whole. The simplicity of the nave contrasts with the rich decoration of the chancel and its elaborate vault. Here there is no central eastern window but a circle of smaller stained glass windows at clerestory level above, and corresponding with the octagonal bays. Behind are the ambulatory and a half-circle of richly decorated chapels – the chevet chapels – which, in their exterior, project to give that distinctive lozenge-shaped appearance.

Although St Patrick's is unashamedly eclectic it is much more than a simple copy of various mediaeval features. There is a unanimity and specific character in Wardell's design - how he has blended two Gothic periods, the simple Early English style of the nave and the Decorated Gothic of the rest of the building; how all the different architectural features are faithful to the general principles of mediaeval Gothic, and how those various elements have come together in perfect accord much in the same way that individual sounds from various instruments in an orchestral work blend to make a pleasing whole. The genius is in the harmony and the proportions of the building that has been authoritatively described as 'the greatest cathedral in Australia'.[19]

We may wonder how Wardell would have reacted to the subsequent reordering of his design – not only the exterior but the concrete extension of the sanctuary into the crossing to accommodate the altar prescribed after the Vatican 11 liturgical changes; the poorly conceived Gothic screen which supports the organ loft in the north transept; and the removal from the crossing of the splendid Gothic pulpit of carved oak installed in 1890. Wardell used to exercise strict control over all aspects of interior design although he lost one argument about the original placing of the organ across the west end of the nave so as to obscure the great Hardman window already mentioned. 'Why put this unsightly structure in front of one of the finest modern productions in stained glass?'[20] he demanded – knowing well that it was his own finest work as well as Hardman's that was disadvantaged. The organ remained there until 1937 when it was moved and it now obscures the window in the north transept.

The problem of where to site the organ gallery so that it does not obscure important structural features – a problem common to both Wardell cathedrals - is perhaps insoluble.

Wardell aged about 40.
Copyright: MDHC Catholic Archdiocese of Melbourne.

CHAPTER 14

UNDER DOCTOR'S ORDERS

Towards the close of 1869, Wardell had completed eleven years service in the Victorian Public Works Department, and throughout all that time he was also managing his busy private practice. By then he was thoroughly bound to his new country, a true Melbournian and part of Victoria's influential establishment. Earlier that year his application for full membership of the Institution of Civil Engineers in London had been accepted. Hitherto he had been a lowly Associate, but now he could add C.E. after his name. This was not mere self aggrandisement; it was part of his counter-attack against those who questioned his competence from time to time. Recognition of status by a respected international body helped to silence criticism and enlarge his field of activity.

But the enormous volume of work before him, both in his Government office and in his private chambers, was bound to take a toll on his health sooner or later. The shadow of Pugin's tragic end, his confinement in a mental institution as a direct result of gross over-work, must have proved an uncomfortable fancy in the background of Wardell's thoughts. There had been so much in the two architects' lives that was common to both. Could not Wardell's health break down as Pugin's had done, so that Wardell might suffer a similar humiliation?

We do not know what symptoms Wardell experienced which led him to seek medical advice but probably they included headaches, inability to concentrate, lassitude and insomnia – classic indications of mental and physical exhaustion.

That he sought medical advice is therefore no surprise - he was probably urged to do so by family and colleagues; but why he chose to consult a phrenologist may seem surprising to us given Wardell's practical-minded rationalism in so many other matters. To understand why he did so we should remember that mid-nineteenth century medicine used a variety of standard therapies which are long ago discredited – for example the application of leeches and the extraction of copious amounts of blood for practically every ailment; the cruel extraction of healthy teeth as a supposed cure for a variety of illnesses, and the opening of the mastoid cavity as a cure for deafness. Like these, phrenology was also widely accepted in the nineteenth century, recognised as a branch of medicine preceding Freud's psychoanalysis and similarly emanating from mid-century Vienna. Phrenology boasted a voluminous bibliography based both on observation and practical experiments and was supported by many reputable scientists although it was never entirely without its opponents.

Briefly, phrenologists believed that they could map the various areas of the brain by careful measurement of each part of the skull, and the contours of the skull signalled the brain configurations. Further, the shape and size of the skull was an indication of the character, the abilities, and even the mental health of the patient. This led to extravagant theories concerning the close relationship between the shape of the head and criminal or deviant tendencies in the poorer classes, and conversely, the abilities, the strengths and the talents of others considered more privileged. Not surprisingly such theories also encouraged a fair degree of lampooning and promoted the doubtful practice of telling fortunes by reading the bumps on the head of the gullible. The only trace of phrenology that survives today is found in the derisive advice directed at someone whom we feel ought to have his head read.

The phrenologist whom Wardell chose to consult was A. S. Hamilton, a well-known practitioner in Melbourne who advertised his services in *The Age*. There he named a fee of five shillings for a verbal consultation only; and for a complete written study including advice for self-improvement for the sum of one pound. His consulting room was located above a jeweller's shop at 76 Bourke Street. Hamilton was evidently an ardent champion of phrenology and a successful entrepreneur for he advertised his willingness to lecture in schools and at private parties 'upon request'.

After producing a detailed chart of measurements of the skull, the report that Hamilton wrote for his patient, Wardell, reads like a Fortune-teller's flattering assessment of character. It tells him what he is most pleased to hear about himself.

'Your brain is large and especially prominent in the intellectual region...
You are exquisitely sensitive both from your cerebral organisation hence
the exhaustive wear and tear resulting from the pressure of circumstances
of a very responsible nature....You are an enthusiast, a devotee, a poet in
feeling and sentiment naturally above all mercenary considerations. You
would desire to treat all subjects on their true merits and would follow art
and science for its own sake.'[1]

Hamilton then describes how three organs of Wardell's brain had
become much depressed and injured by over-exertion 'and nothing short
of rest will be of service to restore the strength and the balance of your
nervous system.' Hamilton followed this report with a letter strongly
prescribing rest, and here he sounds like that ideal doctor we all would like
to consult occasionally in our busy lives, one who will give the advice we
most want to hear: 'You should avoid all care, responsibility and mental
labour. Whatever you do should be simply pleasurable and soothing.
Nothing must be allowed to come between you and your determination
to enjoy rest of body, and rest and happiness of mind.' And he adds:

'You are better alone with a tranquilising book than with companions of
the common mould...and you should avoid sudden changes of prolonged
exposure to cold.' He promises that if Wardell observes his strictures, he
will 'recover, or partially recover your former energy.'[2]

Perhaps Wardell knew beforehand what Hamilton would prescribe
and that his consultation was a way of providing evidence for a period
of sick leave which he guessed he desperately needed, irrespective of a
doctor's confirmation.

Whatever we may think of Hamilton and his report in hindsight,
they were both taken seriously by Wardell's Minister and his colleagues at
the time. A degree of sympathy was expressed and he was granted twelve
months leave from January 1870. Staff in his Department prepared an
ornate illuminated address of farewell signed with evident affection by
twenty-six of his colleagues. This was presented to him on departure on
3 January. In ornate calligraphic script it testifies to 'our deep regret at
the necessity which calls you away from your present sphere of duty and
[we] express our sincere wishes for your speedy and complete restoration
of health.'[3] That he should receive such an elaborate and - we conjecture
- spontaneous testimonial from his Department surely signifies that he

was held in high esteem by his staff and was not the rigid, unsympathetic dictator that his critics and other malcontents tried to paint him.

On 3 January, 1870 Wardell was farewelled by a large party of officials and members of his Department on board the Royal Mail Sailing Ship, *Malta* in Hobson's Bay. There were priest friends there too, and even the Chief Minister, Sir Redmond Berry accompanied by other members of the government.

Presumably his family had made their farewells earlier; they were, by this time, well used to his absences.

The *Malta* had a precious cargo quite apart from Wardell who is our prime interest. On board was Captain Blanchard of the French barque *Marie Gabrielle* which had been wrecked off Cape Otway and was returning to France to face a Board of Enquiry. In the hold of the *Malta* was a rich cargo of gold coin, and notes for the Bank of England. The passengers and crew were not to know that on a later voyage, the *Malta* itself would be wrecked in a storm off the coast of Africa.

The ship sailed through the Suez Canal which had been opened with great ceremony the year before. In keeping with her name, she then headed for Valetta, the port for Malta. Wardell disembarked and inspected the graving dock, making a sketch of it, observing features which he put to good use when supervising the Williamstown dock two years later and which may have influenced his judgement of the Calliope dock in Auckland. Wardell may have left the ship at a port in Italy and journeyed by land through Europe to London because his sketch book depicts small details of buildings and scenes on the Continent that took his interest. In London he stayed at the Stafford Club in Burlington Gardens, a club for Catholic converts founded by Lord Vaux of Harrowden. From there he visited several of his old artist friends, John Rogers Herbert, and David Roberts, and he accepted an invitation to stay one weekend at Edward Wilson's country house in Hayes, Kent. Wilson addresses him as 'My dear old Wardell, from numerous sources I know of your safe arrival in England and I should have called upon you ... but I seldom and most unwillingly come to town now and hear the dreary sounds of the Club members [enquiring] "Not, Sir, too well?'[4] He concludes his letter by encouraging Wardell to visit because his old friend Bellasis will also be there, 'it will be good to have you with me.'[5]

From the limited correspondence and few references to Wardell's
return to London it seems that it was not without some sadness. His
great friend Clarkson Stanfield was dead, and the city and the places he
knew had changed much in the twelve years since his departure. Steel
had replaced iron as the more malleable, lighter but stronger metal, and
its production in ugly smoke-belching steel works in the northern towns
had become the backbone of the Industrial Revolution. London had
become, if anything, more crowded and unhealthy than when Wardell
had left it twelve years previously. He would have experienced the near
impossibility of returning to a well-loved place after a long absence 'for
what we seek is a time as well as a place.'

Wardell would not hanker after his old home in Hampstead. His
home was now in Australia and, in spite of troubles ahead in the Public
Works Department, he must have been grateful towards the end of the
year to board the ship which would take him back to Melbourne and his
family.

CHAPTER 15

MANAGEMENT UNDER FIRE

The great volume of work that had obliged Wardell to take leave for the good of his health showed no sign of abating upon his return to his desk in Melbourne in January 1871. His long sea voyage the previous year and his travels in southern Europe may have provided the rest for his brain that Dr Hamilton had prescribed, but like Pugin before him, Wardell was incapable of reducing his momentum and seemed, on the contrary, to be invigorated by taking on more than the average man could be expected to do. The magnitude and multifarious nature of his duties demanded not only his architectural skill, and the close supervision of every aspect of the work in his department which ranged over the whole of Victoria, but also routine executive administration and staff management. In the Administrative section alone he employed a Secretary, an Accountant, ten clerks and two messengers. In addition to the foregoing he was appointed to a succession of advisory boards often running concurrently. Clearly this was an impossible burden for one man to bear - a fact which the 1873 Royal Commissioners recognised, and recommended that the supervision of clerical duties be separated from the professional division. Some idea of the range and magnitude of professional work supervised by Wardell can by gained from the list of Public Works Department projects he tabled before the Commissioners (cf. Appendix 3).

Among Wardell's ex-officio appointments may be listed his Membership of the Royal Commission on the London International Exhibition (1872), and in August of the same year he was appointed a Member of the Royal Commission into the utilization of low-lying lands on both banks of the Yarra. In the same month he was appointed Member of the Board to examine private bills of the first class. All this involved meetings, reports and research and was over and above his duties as Inspector General of Public Works and Chief Architect. And if we remember that concurrently he was somehow managing to fulfil commissions which came to him in his private practice, and visit regularly the site of St Patrick's Cathedral, we marvel at his work-load which would surely have crushed a lesser spirit. Between 1872 and the crisis in 1878 to be described later, Wardell was appointed a member of two Boards of Enquiry and three Royal Commissions. He occupied the opposite side of the table when he was called as chief witness at the Royal Commission hearings of 1873; and was himself the subject of a Board of Enquiry the following year. As late as March 1876 he was given an additional appointment as Acting Chief Engineer of Water Supply. His dismissal along with other public servants which was to occur in 1878, although grossly unfair, probably saved his life. *The Age* seemed oblivious of these facts when it continued the attack on him and described his position in the Public Works Department as 'a sinecure.'[1]

One of his most important projects on resuming his duties in 1871 was the design and supervision of Government House in Melbourne. Because buildings designed within Wardell's Department were unsigned – deliberately so - we can only speculate to what extent Wardell's ideas are imprinted on them. He had several highly capable architects on his staff, including Samuel Merrett, John James Clark, John Henry Harvey and John H. Marsden, but he maintained a strong supervisory role and set the style and policy to be followed. Describing his general policy to the Commissioners he explained: 'After consideration of fitness and economy ... the simpler the better, so long as proper architectural effect is preserved.'[2] In a memorandum to his staff dated 12 July, 1859, he was more specific.

> 'The gentlemen of the drawing office are requested to be so good as to observe the following suggestions for all buildings in preparation now and in the future:
>
> 1. Parapets are to be avoided as much as possible.
>
> 2. Eaves with a projection as little under 18" as possible.
>
> 3. Eaves, gutters and downpipes are to be invariably of cast iron and galvanised if possible.

4. The brackets for eaves, gutters and fastenings for downpipes to be of wrought iron.'[3]

How much Wardell contributed to designs of public buildings originating within the Department may indeed be a matter of speculation and may never be known, but we do know that he was responsible for the design and general style of Government House because he tells us so himself. In answer to a question put to him by the Royal Commissioners asking whether the plans for Government House were commenced while he was away in England, he answered, 'No, I made the first sketch for a general plan [myself]'.[4] Further questioning revealed that he was involving himself in more detailed planning than was usually the case – the work being far from complete at the time of the Royal Commission.

Designing Victoria's monumental - and Australia's finest - vice-regal residence would have been considered a prized commission. For this reason the Colony's Government decided to make an exception on this occasion by throwing it open to competition among architects in private practice. The winning design was chosen by a Board and paid for and then eventually rejected, which aroused intense ill-feeling on the part of the winning architect, Joseph Reed and his associates. Reed saw Wardell's hand behind the decision - an unfair assumption since Wardell had been a member of the Selection Board that chose it in the first place. Embittered and jealous, Reed became a leading voice in the call for a Royal Commission to enquire into the workings of the Department and, when appointed, was the chief complainant at the hearings.

The Reed designs had been submitted and accepted as early as 1864 but the Government delayed confirming the Board's choice for six years. When Wardell arrived back from England he learned that in his absence the Chief Minister, Mr Bates, had decided Reed's designs 'did not correspond with what was then wanted.' He (Bates) described the plans which were on display at Parliament House, as being 'a rechauffé of the Menzies Hotel [one of Reed's many important buildings], and wished to have a more palatial building in the Italianate style.'[5] Thus Joseph Reed's design had clearly been rejected by the Government and not by Wardell himself. Wardell thought Reed's design 'might be greatly improved [but] it was the best shown.'[6] When Wardell was asked by the Commissioners whether he thought the new design for Government House was better than the one submitted by Reed, he replied modestly, 'I would rather not answer that, for I am responsible for it. I think there is considerably more accommodation in the present house than there was in the other; the bedroom accommodation is better.'[7]

The Royal Commission had been appointed on 6 January 1873 and its purpose was to 'enquire into the system adopted by the Public Works Department, in reference to contractors and the execution of Public Works, and generally to report on the Department itself.'[8] Such terms of reference as these suggest that the Government of Sir Graham Berry was responding to the criticisms of discontented contractors and jealous architects, ably supported by intemperate allegations made against Wardell in *The Age*.

Wardell, as chief witness, was questioned by the Commissioners over the first five days of hearings, and was recalled towards the end of the proceedings for a further two consecutive days to reply to some of the allegations made by critical witnesses. In effect Wardell and his management of the Department were on trial. If so he handled his ordeal with patience, courtesy and precision. He stated in his opening submission that 'The responsibility of everything connected with the Department rests on my shoulders, and although I work with other people's brains as well, the responsibility is mine.'[9]

Among the subjects of interest to the four Commissioners were how tenders were handled and estimates made, and whether the accusations of over-estimating on certain projects were valid; why the designs for the Mint and the Alfred Graving Dock at Williamstown and the Customs House were altered after estimates were made; why so few Government buildings were thrown open to competition and whether competition could be a more efficient, fairer, and in the long run a more economical alternative to in-house designing. They also wanted to know why the foundations of the new Government House were laid before the plans were completed – this last a matter of particular public comment and scorn in newspapers. Wardell was also closely questioned on house-keeping matters: salaries of architects, appointments, and apportioning of duties within the Department, and the amount and variety of work undertaken within the Department.

When asked for his opinion on the subject of competitive designing by architects in private practice, Wardell replied: 'I do not think it is a benefit to the public.' He explains further that although 'it would be a very fair thing for the architects outside', independent architects not attached to the Department cannot possibly know so well as the Department how the public offices are worked and how they ought to be accommodated ... I think you will find that there will be delays and difficulties.'[10] He also points out that estimates based on the competitors' small competition drawings invariably turn out to be far below the total costs once the detailed plans are made. This last point is so often proved true in our own time; the Sydney Opera House a notorious example. Wardell concludes by stating his belief that 'speaking as an architect [my Department] has within quite as good talent as there is outside.'[11]

Turning to the foundations of Government House and why they were laid long before the plans for the building were ready Wardell explained that the Government itself had urged that work be commenced as soon as possible. 'The sum of £15,000 was voted and if we had not taken the contract [for the foundations] it would have lapsed; in addition to which the lease of Toorak [the house where the Governor then resided] expired at the end of 1873 and it was urgent that no time was lost on that account.'[12] In this instance, he said, if they had waited to build the foundations until all the detailed plans for Government House were completed, as was the normal practice with other buildings, the delay in starting work on them would have been excessive.

The Department had also been severely criticised in the newspapers because, according to reports, the foundations had been staked out and excavations had already started on Government House when Wardell decided it was in the wrong place, changing his mind about the location of the building. The newspapers claimed new foundations were then required to be dug, resulting in additional labour costs. 'This was not true', Wardell told the Commissioners. The foundations had merely been staked out based on early sketches but when he had inspected the site he found 'it would be infinitely better to shift the building back.'[13]

No excavation or work was involved in it. Wardell added wearily: 'They say so many untrue things about us that I notice them now very little.'[14]

Joseph Reed appeared before the Commission over two days, on April 23 and again on May 27. In a lengthy opening speech he summarised eloquently and convincingly his main complaint against the Department. Reed was a formidable architect at the time and became a major figure in the story of Melbourne's architecture, a serious rival in stature to Wardell. His most famous and most admired buildings among many others, are the grand, classical Public Library, Rippon Lea House, Victoria Arcade, banks, churches and the Exhibition Building. J. M. Freeland in his History of Australian Architecture states that 'his tremendous abilities enabled him to impress his stamp indelibly on a city which more than any other was moulded by one man. Melbourne, even today, is Joseph Reed's city.'[15] Reassuringly for the biographer, Freeland also adds Wardell to Reed and Blackett, as 'the three giants of Australian Architecture in their time.'[16] The opinions of Joseph Reed, therefore, could not be discounted.

Reed argued cogently that it would be an advantage to get large public works open to competition because 'it would draw the best talent in the colony [and] if a person has an individual interest in a building, he will pay more attention to it than if he were a mere member of a Department, where perhaps, whatever his merits might be, he would not have the individual credit of it.'[17] He then makes the point that the history of nations is written in its buildings and that Melbourne's public buildings have actually suffered from the mere fact of having a Public Works Department. He did not know of another country where a similar system pertained. '... in England all the great works have been done by individual men – Barry, Scott, Street, or someone of their stamp. In France also they are specially works of art and beautiful buildings.'[18]

Reed was put on the spot when he maintained the Department was responsible for building errors and estimates, and some badly designed buildings. The following exchange took place:

Commissioner: 'The outside architects have never been responsible for that?

Reed: 'Not so remarkable as the Public Works Department.

Commissioner: They have all been successful?

Reed: I do not say that they have; they have made some mistakes sometimes.'

Commissioner: 'There have been failures?'

Reed: 'Yes, there have.'

Reed then realises that the questioner is about to allude to one of Reed's own failures, the Bank of New South Wales which, at great cost, needed to be vacated and rebuilt on account of mistakes in the design. Reed blustered and replied: 'One cannot go through life without making mistakes. It does not alter the principle.'[19]
'No', the Commissioner agreed, but then reminded Reed that when the Department makes such mistakes 'it has been reflected on [i.e. criticised] for such architectural abortions.'[20]

Another hostile witness, the architect Thomas Crouch, maintained that a great deal of money had been wasted in building the Williamstown Graving Dock. He quoted clerks who had told him 'the works there have been made extravagantly heavy and unnecessarily thick ... on the principle of making it strong enough whether it was required or not.'[21] In his reply to these accusations Wardell gave the opinion that they were made in ignorance of the technical demands of dock building, and that there were no qualified hydraulics engineers in the Colony - other than those employed within the Department - who understood the requirements.

When the Commission concluded its enquiry in August 1873, among the several recommendations in its Report to Parliament is a suggestion that 'the Government should invite competitive designs for all important State buildings from private architects ... providing [the winner] will consent to work in conjunction with officers appointed by the Department for the protection of State interests.'[22] Accusations, including the charge of waste within the Department, they found were not substantiated by the evidence presented. In the particular case of the Alfred Graving Dock the Commission found that 'it had been absolutely necessary to adopt dressed or tool work to ensure accurate joints and fittings, and to guard against the water finding its way through the stones.'[22] The report stated that there was no unnecessary expenditure and the alteration to the depth of the dock during construction had been ordered by Sir George Verdon, Acting Minister of Works in order to accommodate vessels of greater depth than had been specified in the original Government commission.

In conclusion the Commission avoided attaching any blame for alleged short-comings in administration to the Inspector-General, but recommended that his onerous work-load, as already mentioned, should be reduced. This would allow him to devote more time to his architectural responsibilities. In the public eye, Wardell was exonerated.

Wardell's ordeal in front of hostile questioners was not over with the conclusion of the 1873 Royal Commission. A little over twelve months later he was required to face, not another Royal Commission, but a Parliamentary Board of Enquiry. This had been appointed to investigate the truth of charges brought against him by a member of his Department, Thomas Andrew Eaton. Eaton claimed that Wardell was prejudiced against him and that he was thereby unfairly excluded from the supervisory works which it was his duty to superintend; that he was denied promotion and was left with very little work to do. Eaton had published a pamphlet detailing his complaints and listing what he maintained was evidence of their validity.

Thomas Andrew Eaton's position in the Department was Travelling Superintendent Inspector of Works for the Metropolitan District at a salary of £485 per year. The normal annual increments from this point, he claimed, had been blocked by Wardell and he had been given less and less work until the amount was 'ultimately reduced to nothing'.[23]

The dispute dated back to 1866 when faulty work at the Kew Lunatic Asylum was under investigation and Eaton had refused to inspect the building, believing further examination was unnecessary. Another cause for dissension was Eaton's refusal to sign a petition to the Government requesting Departmental Officers be allowed to engage in private practice. Eaton claimed that Wardell had said to him: 'I shall remember your ingratitude.'[24]

Yet another reason cited by Eaton for Wardell's animosity was supposedly religious prejudice. Eaton was a Protestant, believed by Wardell to be an Orangeman.

Over the first three days of the Enquiry Eaton, accompanied by his solicitor, Mr Gaunson, was subjected to close, detailed questioning. He called witnesses to substantiate his claim that Wardell was prejudiced against him and that he had been badly treated. But the report on the dispute reminds readers that 'a grievance can distort the facts'.[25]

Wardell denied each of the charges both in a written statement and during his appearance before the Board. On the question of denying Eaton promotion because he was a Protestant, Wardell replied:

> 'It is humiliating to have to answer such statements and I would rather refer to the experience of every officer with whom I have worked for the past fifteen years as to whether my dealings with any of them give colour to such an implication ... If any such thing as that existed on my part, it would, I think, have been manifest in other cases, and probably have restrained me in the course I took in Mr Eaton's case on the two occasions on which he had been promoted. I have never had any such feeling; but if I had, I hardly think I should have been foolish enough to assert it to a reigning Government [Representative] as a reason for not complying with his request for the advancement for anyone he thought deserving.'[26]

Wardell admits that he felt it his duty to give Eaton as little work as possible. 'Mr Eaton is an elderly man, and extreme measures have been avoided by the consideration of Ministers, and I think it will be seen that to be relieved of work he should have done, without inflicting any loss on him, is hardly a persecution.'[27]

Wardell in his turn referred to counter-charges against Eaton but denied that they were charges made solely by him. '... they are in fact, reports by certain Boards appointed to enquire into certain works superintended by Mr Eaton ... reports unfavourable to him.'[28] It was on account of these reports that Eaton had been relieved of many of his duties.

The Enquiry Board's Report to Parliament was dated November 25, but its conclusions which exonerated Wardell of prejudicial behaviour towards Eaton were made public as early as October 21. *The Advocate*, in noting that the Board had concluded its investigations, reported that, '... so far as regards the charges preferred by Mr Eaton in his pamphlet against Mr Wardell, [it] decided the charges were not proved.'[29]

There was, however, a mild rebuke in the final paragraph of the

Report about the lack of memoranda in the Public Works Department in connection with Wardell's 'grave act...of allowing a public officer, whose services he declined to use, to draw his salary without giving an adequate return to the State.'[30]

Wardell's humanitarian gesture on this occasion was not appreciated by his interrogators.

CHAPTER 16

A GREAT VOLUME OF WORK

That indefatigable traveller and commentator, Anthony Trollope, was on his visit to Melbourne in 1873, at a time when early construction work on Wardell's Government House was clearly visible. As an important visitor already well known at the time for his Barchester novels, he would have been taken to see the site and shown the plans. Most likely Wardell, remembering his old literary contacts, had taken him there himself. Like countless other visiting artists before and since, Trollope probably kept any criticisms he may have harboured to himself. Later, he would write, in his celebrated account of his travels in Australia and New Zealand, his opinion that the new Government House was going to be too magnificent and too expensive to maintain, 'Were I appointed governor of a colony I should deprecate very much a too palatial residence. I think it may be admitted as a rule that governors find it hard to live upon the salaries allotted to them, and generally do not do so.'[1] - a shrewd judgement which was remarkably prescient in view of the secret salary arrangement negotiated between the reigning governor, Sir George Bowen, and the Premier, Sir Graham Berry, related in the next chapter.

Trollope's use of the word magnificent to describe Government House (though surely not 'too magnificent'), was far from exaggeration. Magnificence suggests grandeur. And it is certainly grand. When the visitor, approaching the mansion from the drive, first sees its cream stucco and classical design caught by the late afternoon sun, there is revealed a grandeur as well as stately confidence and astonishing beauty unequalled by any other vice-Regal residence in Australia.[2] Melbourne's Government House represents a break in Wardell's hitherto almost exclusive Gothic work, and shows that he is also a master of Classical, Italianate design.[3]

It is commonly thought that Government House, Melbourne, was inspired by Prince Albert's Osborne House on the Isle of Wight (*c*.1847), and certainly the stately belvedere tower with the Vice-Regal standard flying from the masthead which appears above the trees in both instances lends support to this view.[4] But the similarity is only superficial. Wardell's design, made fresh after his return from a visit to Italy during his European tour and mindful of his Minister's instruction calling for the Italianate style, is more likely to have been heavily influenced by Italian Renaissance palaces such as the Villa Farnesina in Rome, or even the Gran Guardia Vecchia in Verona which, like the Ballroom and Reception Wing of Government House, features semi-circular arches on the ground floor and a piano nobile above. The tower belvedere is similar to the Campidoglio by Michaelangelo in Rome. Whatever the influence may have been, this is no tired copy. The result displays originality and a distinctive character, a magnificent Palladian mansion with a classical interior which would have horrified his mentor, Pugin. But Wardell was, to repeat his grandson's words, 'his own man'.

The plan view of the building shows three distinct yet cleverly integrated wings unified by the elegant Tower, a handsome landmark already mentioned. When the elevation is viewed from the gently sloping West Lawn, the East Wing of two stories is on the right, comprising the magnificent Ballroom surrounded by extensive terraces, gardens and lawns. The State Apartments rise three stories in the centre of the picture flanked by the Tower. Here are located the State Dining-room, the grandly proportioned State Hall with its free-standing Corinthian Columns and cross-vaulted ceiling; the sumptuous State Drawing Room leading through an archway to the Conservatory beyond. The North Wing on the left of the picture, like its opposite Wing on the right, is of two stories, and comprises the private Vice-Regal apartments, the administration offices, the Governor's Study, Private Dining-Room, Drawing Room and Morning Room - all these no less elegant than the other parts of the house. The facade of the principal block has six bays, the pedimented windows of the first - central - floor being larger than those below and above, indicating the importance of reception and ceremonial rooms. The hipped roof is concealed by a balustraded parapet which is continued around the north and south wings and helps to unify the three sections. The genius is mostly noticeable in the elegant proportions of the rooms and the restrained decoration mostly confined to the coffered ceilings with their richly formed mouldings and low reliefs.[5] The Ballroom was said to have been the largest in the British Empire at that time, larger even than the Ballroom at Buckingham Palace. Five years elapsed between the first drawing of the plans and the grand opening of Government House marked by a ball to celebrate Queen Victoria's Birthday in 1876. Governor Sir George Bowen, reporting on the occasion to the Secretary of State for the Colonies, wrote, 'The accommodation was so good that there was no crowding, although nearly fourteen hundred persons were present. The toast of Her Majesty's health proposed by me at the Supper, was received with great enthusiasm.'[6] Among the guests on that occasion were Mr and Mrs William Wardell.

Government House, Melbourne, tells us much about the character of the architect - more perhaps, than is revealed in his Gothic work. For here in this building - his one palatial, classical residence - we see nobility, strength yet restraint, elegance, discipline, and informed taste, attributes which we may apply to Wardell himself. Another attribute which visitors do not see because it lies hidden underground, is the water supply, a masterpiece of hydraulic engineering. Huge tanks under the stable yard at the back of the building conserve all the rain-water from the roof.

Before his Government House was completed, and immediately after the 1874 Board of Enquiry into the Eaton affair, Wardell was to make another sea voyage. His reputation as an engineer with experience of harbour re-construction had been noted in distant Western Australia. At that time the Colony was still twelve years away from independent elected government, but chief among its pressing needs before the Executive Council was the on-going question of where to site the colony's chief port. The anchorage off Fremantle was notoriously hazardous, most famously described by an American sea-captain at the time:

'It is a terrible place. No place to put a vessel. No shelter whatever It is certainly the worst place I or anyone else ever saw. And any man who would come or send a ship a second time is a damned ass.'[7]

With comments like these, and other less colourful, but no less critical reports, the hopes and future advancement of the Colony were largely dependent upon the provision of a safe and convenient port capable of berthing the largest vessels close to the centre of main population. Wardell would be the first of several engineers whose advice was sought on this seemingly intractable problem. Later would follow the British harbour engineer, Sir John Coode, and finally C. Y. O'Connor. Earlier Sir John Coode, without visiting the colony, made recommendations based on Admiralty Marine Surveys although neither of his two suggestions was adopted at that time. Now it was Wardell's turn to offer a solution.

The request for Wardell's services was signed by the Governor in Council in the West and when received by the authorities in Victoria ensured that he would be granted three months leave of absence. He sailed for Western Australia on the mail steamer *Pera* leaving Melbourne on 12 November, 1874.

The ship berthed at King George's Sound (Albany) which, because of its superb land-locked, safe anchorage, was the preferred port of mail steamers calling at the colony. Wardell was impressed and described it in his report as 'one of the finest harbours for its area in the world.'[8]

Wardell spent two months in Western Australia. He consulted widely and considered the various schemes that had been put forward by experts and others for securing the port at Fremantle; and he made his own observations. It is significant, in view of what C. Y. O'Connor achieved some sixteen years later, that Wardell rejected the possibility of blasting away the rock which blocked safe passage at the Swan River entrance, and thus building the harbour at the river's mouth. He accepted the common belief at that time, which O'Connor was to question and reject, that sand drift from the north-west would silt up any attempt to build breakwaters or moles projecting from the shore.

> 'Any money spent in opening a larger channel by blasting the rock bar or otherwise, would be thrown away, for a sand bar would inevitably follow. There is already a sand bar outside the rock bar and with a rise and fall of tide so limited as it is here, it is hopeless to attempt any improvement at the river's mouth.'[9]

Wardell then concludes '... after due and careful consideration of the various schemes,' that the only solution would be the construction of a breakwater to the west of Gage Roads, well outside the river entrance.

The colossal cost, he estimated, would be between £800,000 and £1,000,000. He adds, almost as an addendum, that although he has a 'great desire to avoid any expression which might appear to suggest any question of policy,'[10] he felt it his duty to express the desire that the Government consider the cheaper and his much preferred alternative of making Albany on the south coast the chief port. He states his opinion that,

> 'there would be better advantages to the general interests to the colony by connecting one of the finest harbours in the world by means of a railway from Perth and Fremantle. The harbour at King George's Sound will require no expenditure for a breakwater or any other artificial protection...'[11]

The Report on Fremantle Harbor, Western Australia, and Proposed Improvements, by W. W. Wardell Esq., C. E. is dated 18 January, 1875 at the Government offices in St George's Terrace, Perth, where Wardell was temporarily accommodated. Three days earlier he had been presented to the newly appointed Governor, Sir William Robinson, at a levée at Government House.

Western Australia had to wait another sixteen years and the arrival of the engineer C. Y. O'Connor before a decision was made for a fine deep water port at the mouth of the Swan River which has served the State well until this day.

In hindsight Wardell, in his role as Chief Engineer of Water Supply, was more successful in his judgement on proposed improvements to the port of Warrnambool on the south coast of Victoria west of Geelong. In this instance his design to shelter the anchorage called for a projecting mole of *pierre perdue,* a scheme he had rejected at Fremantle. The cost was significantly lower than the cost of a later scheme put forward by the eminent English harbour engineer, Sir John Coode. After Wardell had left Melbourne to live in Sydney a newspaper report suggested a direct conflict between Coode's report and Wardell's, implying Public Works Department incompetence. Wardell wrote to the *Argus* defending his reputation and that of his old Department:

> 'The instructions given to Sir John Coode were to design "the best means of improving the shipping facilities at the port", and he has no doubt designed a perfect harbour at the cost of about £300,000. On the other

hand I was directed to provide means of sheltering the anchorage and the existing jetties to as great an extent as possible at the cost of about £30,000. It is gratifying to find that Sir John Coode adopts the site I proposed, with the exception of a slight deviation. Without questioning, therefore, the soundness of Sir John Coode's report founded on the facts before him, I am confident that the breakwater designed by me [is satisfactory] for the purpose for which it was intended.'[12]

Safely returned once more to his desk in Melbourne and again in command, and having an enhanced reputation from his consultancy, it might be assumed that William Wilkinson Wardell's position as a senior member of the Victorian Public Service was not only highly valued by the government, but unassailable. Had he not faced two searching Parliamentary investigations - the Royal Commission and the Board of Enquiry – and survived with his reputation and competence undamaged? But the storm clouds were gathering on the horizon. His critics were not silenced and a recent change of Government led by Sir Graham Berry – 'a fiery political radical' - was less inclined to support him. The permission for Public Works Department employees to engage in private practice, which had been defined by the Commissioner of Public Works in 1861, was now rescinded by Cabinet, 'due primarily to strained, personal relations between the Berry Government and himself.'[13] Wardell's final three years in Melbourne register no new commissions for churches.

Wardell's largest parish church in Melbourne, and the most impressive in Victoria, is St Ignatius, sited at the crest of the hill in Church Street, Richmond. Its tall spire, added later by architects Connolly and Vanheems, is prominent, a well-recognised landmark to the north-east of the city. Wardell's interior plan shows a passing similarity with the Cathedral – the Cathedral on a smaller scale - but surprisingly vast for a suburban parish church. The nave of six bays is Early English with characteristically minimal decoration; and a chancel arch and apse with three chevet chapels radiating from the five-sided ambulatory. Like St Patrick's, it has five mullioned stained glass lancets (St Patrick's has nine) set at clerestory level above each bay behind the high altar. There is a beauty in the cathedral-like proportions here and a similar dignity and unfussiness. The slender piers and clustered arches emphasise and exaggerate the great height of the nave, twenty-seven metres, close to ninety feet from the floor to the roof ridge. The width, including the aisles, is twenty-one metres, near seventy feet, and together with the crossing gives a general feeling of spaciousness which we also get in St Patrick's. The visitor who is fortunate enough to have also visited Wardell's so-named 'Cathedral of the East End' in the Commercial Road, London, may suppose that St Patrick's is another cathedral of the East End, but the east end of Melbourne.

The foundation stone of St Ignatius's was laid by Archbishop Polding on 4 August, 1867 and thus Wardell's engagement as architect was within eight years of his arrival in Melbourne. The Jesuit in charge at the time was Father Joseph Dalton who would later establish Riverview College in Sydney and again engage the services of Wardell - although with a less happy consequence. Building work and money raising occupied most of three years before the nave of St Ignatius's was sufficiently advanced to be available for church services. At the Grand Opening a Solemn Mass was celebrated with pomp and ceremony early in 1870 while Wardell himself was sailing overseas. His two sons, Edward aged twenty at the time, and Bernard aged twenty-two, represented him. As part of the ceremony the sons presented the architect's plans of the church to the Rector, Father Dalton. On the occasion of the Mass, a parishioner, Peter Lalor, the famous leader of the Eureka Rebellion, took up the collection. Lalor was then a respected Member of Parliament.[14] Sadly almost nothing remains of the 'puginesque' stencil decoration. Gone too, are the suspended rood screen and the parclose screen from the choir chamber.

It will be remembered that Edward and Bernard were the two eldest sons who had been left behind in England to finish their schooling at Stonyhurst. Both now in Australia, Edward would later become Director of the Royal Mint in Melbourne. Bernard would be employed as a draughtsman in Wardell's private practice, where he quarrelled with his father and in anger left the parental home and disappeared overseas. The details are sketchy but it is claimed he was unstable and wayward, and believed that he should have been given more credit for his work in his father's office, particularly in respect of the St Ignatius plans.[13] Wardell was undoubtedly an authority figure within the family, strict and demanding, but not more so than was the custom of the age. His fatherly care for his children, as remembered by his grandchildren, was said to be kindly, and the break with Bernard would have distressed him.[16]

On 27 September, 1873 Wardell celebrated his fiftieth birthday. A studio photograph of him taken around that time shows him fashionably dressed in a frock coat with a cane in his right hand and with his left elbow resting casually on a tall three-legged table. He is evidently prosperous. His watch and chain dangle visibly over his lapel. He is heavily bearded, grey, and his ample moustache droops either side of his mouth. His remaining hair, though thick, is bunched either side of his head above the ears - he had a receding hair line and a high, balding forehead since his early thirties. This and a second, closer portrait taken about the same time shows a determined, character-full face, but the eyes are kindly, and display a hint of humour, as though he is amused by the studied seriousness of the occasion. He seems an amiable man in spite of his air of importance.

The figure we see in the portraits is relaxed and confident of his place in society, seemingly undisturbed by the slanderous accusations that were constantly levelled against him in the press - notably *The Age*. Wardell had his persistent detractors. Some critics were motivated by professional jealousy, others resented his power and patronage in the Public Service, and not a few, as we have seen, were motivated by anti-Catholicism.

Whether the Government Leader, Graham Berry, was moved by these opinions to sack Wardell or influenced by them, we cannot be sure, but according to one authority, 'There is little doubt that ... the Berry Ministry was anxious to rid itself of the services of Wardell in particular, because of personal feelings and the general tendency of that Ministry to fill appointments from its political supporters.'[17]

Top positions in Australia's State Public Services have always been subject in some degree to Government interference; but the crisis that was to come was unprecedented and was probably illegal. Although not specifically directed at Wardell, Wardell was the biggest catch when Berry cast his net.

CHAPTER 17

BLACK WEDNESDAY AND DEPARTURE

Wardell had been resident in Melbourne and working in the Public Service for twenty years when the Constitutional Crisis came to a head in early January 1878.

In an uncanny way the circumstances bear some similarity to the constitutional crisis leading to the Dismissal episode of 1976. Both arose from a long-running squabble between the Upper and Lower Houses, the one in State Parliament in Victoria - our immediate concern; and the other almost a hundred years later in the Federal Parliament in Canberra.

The Victorian Assembly led by Premier Sir Graham Berry had been re-elected in May 1877 with a comfortable majority. It had promised trade protection, land tax reform and reform of the Legislative Council. The issue which brought the two Legislatures to the point of crisis was the introduction of a Bill to pay members of Parliament - a principle widely recognised by democratic movements in all countries and one of the aims of the Chartists. The Government argued that a salary 'was essential if men from the lower-socio-economic levels were to be able to stand for Parliament or were to be represented by men of their own kind. Anti-democrats felt that only men of property and education should sit in Parliament and sneered at the thought of boot-makers and engine drivers sitting among their social betters in the nation's legislature.'[1]

Adopting similar stalling tactics as those used by the Senate in the Federal Parliament of 1976, the Appropriations Bills were repeatedly blocked in the State Upper House, resulting in Parliament adjourning for the 1877 Christmas recess without Treasury receiving the routine authority to expend necessary funds.

Using the intransigence of the Council as an excuse, Sir Graham Berry retaliated harshly after the Christmas break. On Tuesday 8 January 1878 he announced in the *Government Gazette* a list of cost-cutting measures involving dismissals of senior members of the Public Service, including judges, magistrates, coroners and others; prominent among them the Inspector General of Public Works, William Wilkinson Wardell. Some of the dismissals had to be cancelled two weeks later, notably those connected with the administration of law and order, but Wardell was not named among those re-instated.[2]

The *Argus*, invariably critical of the Berry Government, reacted immediately in support of four heads of Departments, Thomas Higinbotham, Byron Moore, Labertouche, and Wardell. In a combative editorial which later provoked a reactionary outburst from the *Age*, the paper declared, 'Each of these officials has occupied a high position with credit to himself and advantage to the country. Each has the reputation of having fulfilled the responsible duties devolving on him with ability, fidelity, and unblemished integrity. Each was justified in considering his appointment a permanent one; and of course laid out his plans in life, adjusted his expenditure and made his social and domestic arrangements accordingly.'[3] The *Argus* then went on to remind readers of the Civil Service Act which all but guarantees permanency of tenure to those of the 'highest attainable capacity and probity.'[4]

This was too much for the *Age* which retaliated, showing no sympathy for the public servants who were dismissed. An open letter even admonished the Governor, Sir George Bowen, not for acquiescing in the order but for expressing regret at the sackings. In accordance with the Constitution it was, of course, necessary for the Governor to sign the order for the dismissals. That he did so with reluctance, as he claimed, and after taking legal advice, did not absolve him from attack from both sides – supporters of the Government action and the Opposition. The *Age* attack was in the form of a letter published three days after 'Black Wednesday'- a description coined by the *Bulletin* - and was signed enigmatically 'One of the men of the Mountain'. Judged by the prominence given to it and its considerable length, the writing had the unmistakable weight of an editorial. If not written by the editor it clearly reflected the views of the owners of the newspaper:

> 'It may perhaps occur to you [Sir George], in mitigation of unbridled grief, that the gentlemen whose services have been dispensed with have for many years been drawing salaries equal to the income of a thriving merchant,

and that – unless they have lived from hand to mouth, as though the country was theirs and the fullness thereof – they have no considerable sum saved for such a rainy day as this has been. It is true that the *Argus* – insatiate for blunders when it would serve a friend – naively says that the civil servants imagined that their billets were for life, and "made their social arrangements" accordingly. Just so. A "Government Billet of £1000 a year was a life-long provision. Henceforth the happy recipient toiled not nor spun. He walked to his office at ten a.m. and he left it punctually at four p.m. He lived in South Yarra, was a member of the Melbourne Club, and occupied a graceful leisure by writing paragraphs for the *Argus*."[5]

These last arrows seemed aimed specifically, yet grossly unfairly, at Wardell. In his Memoir his grandson, Vincent Wardell, adds that there was little doubt that the Berry ministry was anxious to rid itself of the services of Wardell in particular 'because of personal feelings and the general tendency to fill appointments from its political supporters.'[6] The Cyclopaedia of Victoria, in its reference to Black Wednesday states that Berry 'publicly acknowledged that his motive was revenge.'[7]

Historically this proved a shameful episode in Victorian politics which earned lasting censure, much of it directed at the Governor himself for being, as expressed in the *Argus*, a 'consenting party' to the 'coup d'état'; and accused him of 'quasi-criminal weakness.' His reputation was not improved when it was revealed later by the President of the Legislative Council that Bowen had bargained with the Premier for an increase of £2000 per annum in his salary and allowances prior to the signing of the order – vindicating Trollope's opinion of governors' salaries expressed five years previously. A close alliance evidently existed between the Governor and Berry, supported by confidential letters passing between them – one marked, 'Do not mention this to anyone. Burn this letter.'[8] It is hard to avoid the conclusion that Berry bought Bowen's acquiescence. Sir George Bowen, one time Governor of both Queensland and New Zealand left Victoria the following year 'under a cloud'. And inevitably his next posting, to the island of Mauritius, could be seen as a demotion.

In practical terms Wardell, like the other leading civil servants, was 'sacked' - an ambivalent word that is used in some short accounts of Wardell's life omitting an explanation of the circumstances. The wholesale dismissal of more than fifty senior officers from the Crown Lands and

Survey Department, twenty-four from the Department of Mines, and fifteen from the Public Works Department, and many others, all for political reasons questionable in law, hardly warrants the imputation of impropriety attached to the word sacking. The Melbourne Archdiocesan weekly, the *Advocate,* in reporting the event, described it as the 'Commencement of the Revolution.'[9] And it must have seemed nothing less than a revolution to the people who read of Berry's extraordinary action in their newspapers on the morning after the dismissals were promulgated in the Government *Gazette.*

Wardell's salary of £1000 per annum was suddenly terminated. He had been for some time deprived of his right to private practice. It is unlikely that he was without considerable capital, but the demands made upon him as the head of a large family were heavy and required that he continue working. At the age of fifty-five he was too young to retire even if he had wished to do so; architecture was his life.

In those four months immediately following the dismissals and before his departure from Melbourne, Wardell must have discussed with himself, and surely with his family, the options for his future. He could have stayed on in Melbourne and built up his neglected private practice. At the Royal Commission hearings Wardell had claimed that he had, 'as a matter of course, an *ésprit de corps* with my professional brethren outside [the Department]', but he had rivals there, notably Joseph Reed and others who bore a grudge against him. It would have been difficult for him to function profitably having the stigma of the dismissal upon him - however undeserved that was.

Returning to England was never a practical option. After twenty years absence, many of his old friends, associates and patrons had died, including the ones who had given him glowing testimonies when he left the country - Stanfield, Hope-Scott, Lord Petrie, Edward Bellasis, Cardinal Wiseman, and others of his circle including the Earl of Shrewsbury and Pugin. A new generation of Gothic Revival church architects, George Scott, Butterfield, Street, and the Catholics Hansom and Goldie, were dominant and well-established figures. Wardell, at the age of fifty-five, was not someone who could start from the beginning again. But in Australia he could count on his solid reputation and important contacts from as far away as New South Wales, and this encouraged him to look in that direction. Progress on his St Mary's Cathedral in Sydney had slowed due to lack of funds but a new archbishop, Roger Bede Vaughan, had succeeded Polding and had aroused new enthusiasm for the project. Vaughan was determined to move towards completion; the walls of the northern section had reached a height of thirty-nine feet by 1877 but income from donations had fallen in the year 1874-75 to £4195. In the following year, due largely to Vaughan's enthusiasm and commitment, income had doubled this figure which enabled the third-stage contract to be let.

In February after the dismissal, Wardell made an exploratory visit to Sydney, renewed contacts there, and visited the site of the Cathedral. What he saw, and what he experienced in the city on that visit, decided him that his best option was to take himself and his family out of Melbourne and head north for New South Wales.

TOP: The architect's great desk with the drawing board top raised. Designed by Wardell and shipped out from London. It now rests in the Wardell Room at the Melbourne Archdiocesan Historical Commission Museum.

LEFT: The 17th Century rosewood Spanish Crucifix belonging to Wardell. Behind are portraits of Wardell's mentor, Pugin, the original by Wardell's friend, John Rogers Herbert. Beside it is Wardell's photograph probably dating from around 1875.

186

St Ignatius, Richmond. Wardell's largest parish church in Australia. The nave, aisles, and part of the tower were opened in 1870. The blessing of the completed church took place in March, 1894.

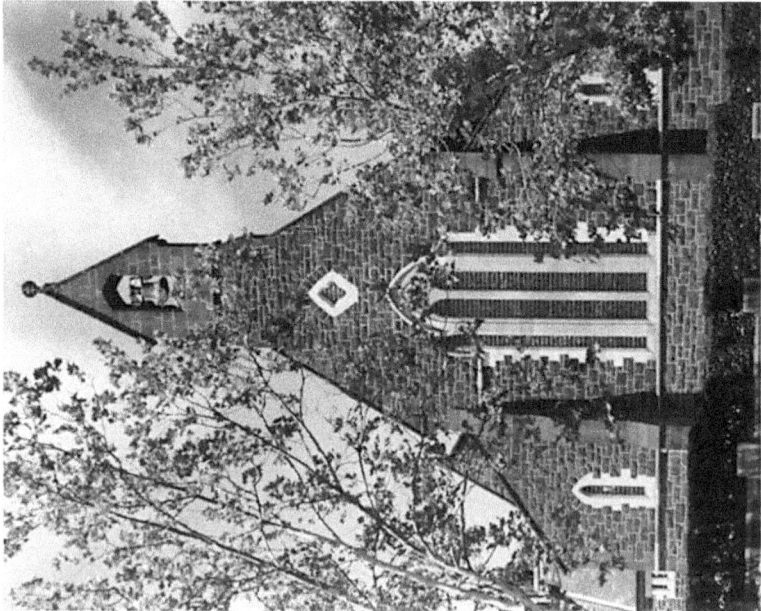

St Mary's, Dandenong Road, East St Kilda. Wardell's original church of 1858 was much smaller. The larger church seen here is the second design built in 1870 and is the most perfect example of Wardell's modest parish churches in Australia.

Two views of Government House, Melbourne, the largest and most palatial Vice-Regal residence in Australia. Designed, and building supervised by Wardell 1871-1875.
Copyright: James Renner and Michael Ritchie, published in the Government House guide book (1994) by the Office of the Governor.

Two views of St John's Heidleberg. (1859). The pinnacles arising from the diagonal buttresses either side of the west front are a unique and attractive feature. The porch and west front were completed in 1891.

St Peter and Paul's Church, Ashby, now known as West Geelong. (1861). Note the unique double bellcote in the exterior photograph. The arches in the chapel on the left (not visible) are of unequal dimensions giving a lop-sided aspect. A builder's blunder which angered Wardell.

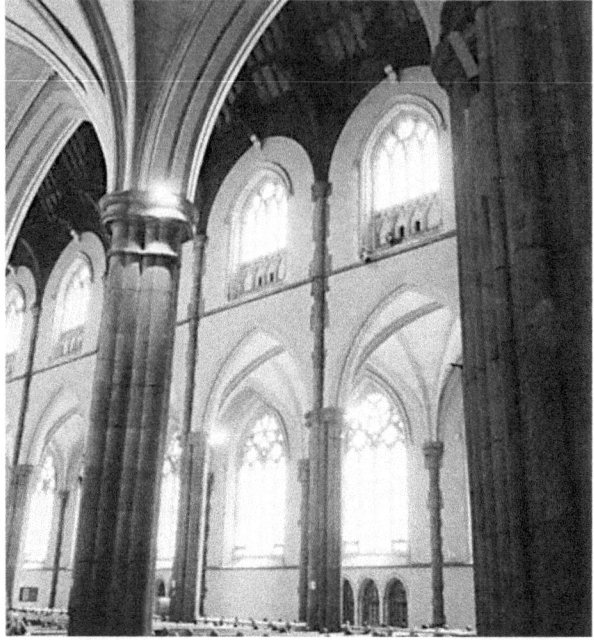

RIGHT: Looking across the nave of St Patrick's Cathedral. Copyright, Melbourne Archdiocese Historical Commission.

Wardell's 'Gothic Bank' located on the corner of Queen and Collins Streets, Melbourne. It was voted by readers of 'The Age' newspaper to be their favourite building in the city.

The interior of Wardell's 'The Gothic Bank' showing the coffered ceiling, cast iron pillars, and ornate arches.

Two of Wardell's suburban E.S.&A. Banks now converted into private residences. (Top): On a busy intersection in Moonee Ponds, the elevation skilfully follows the angle of the corner. (Below): A simple but elegant classical-style building in North Melbourne.

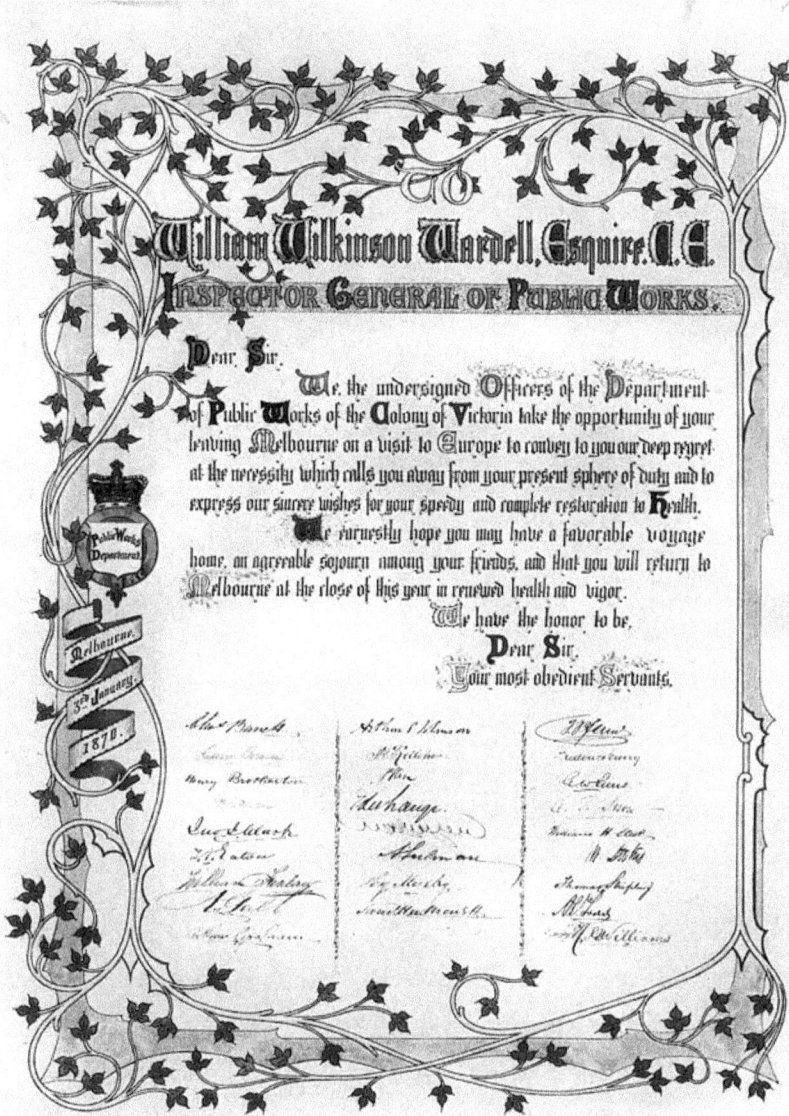

The decorative memorial prepared by his colleagues, and presented to Wardell on his departure on sick leave for Europe in 1870. Among the signatures is that of Thomas Eaton who would later accuse Wardell of unfair treatment. Copyright: Mitchell Library

The great East window of St Patrick's Cathedral, designed by Wardell and constructed by John Hardman in Birmingham.

TOP: St Patrick's Cathedral. The upper sanctuary wall and apse windows beyond, showing the rich stencilling and graceful arches.
(Picture by Jane Pertile©)

LEFT: St Patrick's Cathedral.
The juxtaposition of soaring columns viewed from the side aisle and the vaulted stone ceiling are important elements of Gothic design intended to raise both the eye and the spirit heavenwards.
(Picture by Jane Pertile©)

TOP: St Patrick's Cathedral: The apse exterior showing the pinnacles, crockets and finials that adorn the chevet chapel roofs, delighting the eye with their visual complexity.
(Picture by Jane Pertile©)

LEFT: St Patrick's Cathedral.
St Patrick's Cathedral. General view of approach to the east transept.
(Picture by Jane Pertile©)

198

SYDNEY

1878 – 1899

CHAPTER 18

'NEW SOUTH WALES ENRICHED'

Wardell had only recently moved from his long time residence, Ardoch, in Dandenong Road, East St Kilda, to reside in the more fashionable Toorak Road, South Yarra. That house which he and his family occupied for only about a year, no longer exists. Hardly had they settled there before Wardell, in the aftermath of 'Black Wednesday,' up-rooted again to take his family to New South Wales, departing finally in mid May 1878. He was not turning his back on a city which he had helped to build. Greatly wronged as he was by the Berry Government and defamed in certain sections of the press, he seemed - to judge from his occasional writings - remarkably free of rancour. He was not shaking the (Melbourne) dust from his feet as the Disciples were advised to do or 'fleeing from one town to another',[1] although surely he could have been forgiven had he done so. As the Catholic *Tablet*, noting his departure, reported at the time:

> '[His departure] will be good news for those who have been persecuting him with a bitter hatred for many years but who could never gain their object ... He was subjected to the most mendacious and unscrupulous attacks ever levelled against the head of any department ... The more wild and more fierce the assault, the more convincing his refutation and the more signal his triumph.'[2]

The *Sydney Morning Herald*, appealing to a more general readership, was equally scathing in its condemnation of Berry's politics and welcomed the prospect of Wardell's arrival in Sydney.

> 'It is an ill wind that blows nobody any good. The wholesale dismissal of public officers, which has been shown to be unwarranted by any commercial necessity ... has forced several practised officials to look around for a living, and what Victoria loses, other colonies may gain.

New South Wales, it appears, is to be enriched by the possession of Mr Wardell.'[3]

Closely following this very public misfortune affecting the Wardell family, came a very private sorrow. The fourth son, and the second to die prematurely, was Francis William, aged twenty-two. The death notice in the *Advocate* appeared on January 21, just two weeks after Black Wednesday.

Although from this time forward the family would make its permanent home in North Sydney, Wardell himself would be back and forth between the two capitals for the rest of his life. He would continue to involve himself in the design and furnishings of St Patrick's Cathedral as well as to undertake further commissions in Melbourne. Wardell saw the abrupt end of his Public Service career as an opportunity for a new beginning elsewhere, an opportunity to build up a private practice free of Public Service restraint. Unrealised by the Government (and probably by Wardell himself) the dismissal may have saved his life. With his history of fragile health he could not have continued for much longer at the level of activity he espoused. The break came at the eleventh hour.

Maintaining a thriving practice meant soliciting commissions as he had done after arriving in Melbourne. One of the first letters he wrote in advance of his exploratory visit to Sydney in February that year, was to the Building Committee of St Mary's Cathedral where work had temporarily slowed to a halt for lack of funds. 'It is probable that at this stage,' he wrote, 'I may be able to give some suggestions and advice that may be useful and economical.'[4] And then he added that although his normal fee for this service would be fifty guineas he was prepared to accept half this figure including expenses because 'I may be able to make use of my visit for other purposes.'[5]

It is important in the understanding of Wardell's character to accept that he was meticulous in stating and then demanding his professional, standard fees. As much as we may like to think that someone so loyal and so spiritually committed to the Church would tend to be careless of profit and waive his fees in deserving cases, we observe that although there are some exceptions recorded, these were few. He worked according to the rules of his profession and maintained a rigid sense of professional etiquette. He had the reputation within the family of being very careful with money. He would account for every penny, insistent upon claiming what was legally due to him. Constantly throughout his professional life we find in the records that he battled to obtain fees which he believed were due to him, and we sometimes wonder at his persistence. The struggle for settlement connected with St John's College is a good example. This attitude was an element in the Victorian character, a respect for money honestly earned and prudently used. 'A Penny saved is a Penny gained', is a Victorian adage and Mr Micawber's advice to David Copperfield would have been taken to heart by Wardell. On the other hand we find some evidence that he could be excessively generous, and, having insisted upon fees, would donate expensive church furnishings and occasionally his services. There is also evidence that he gave generously to charity.[6]

After initial hesitation on the part of the Committee, Wardell's offer of advice leading to the regular building inspection of St Mary's was later accepted. When the Committee demurred he wrote that during his February visit the Archbishop himself had suggested to him, 'of course in a friendly way,' that he should settle in Sydney permanently.[7] This, he argued, helped him to decide what to do when he returned on that occasion to Melbourne.

And so began, or more accurately was resumed, his intimate, almost daily involvement with the building operations on St Mary's right up until the time of his death. He was fortunate in that his arrival in Sydney coincided with the determination of the second Archbishop of Sydney, Roger Bede Vaughan, to complete the Cathedral in spite of inheriting crippling diocesan debt and finances 'diabolically mismanaged'. Polding, old and feeble, had died the previous year, and had, in Vaughan's words, 'let things run terribly to seed and disorder'.[8] The first contract for work on the Cathedral had been completed by the builder, John Young. Henceforth it would take a leader of exceptional courage and imagination to commit the poor diocese to the next stage contract amounting to £21,000, more than double the previous one. Fortunately the leader with those qualities was Vaughan. When he addressed the Building Fund Committee on September 17 of the year that Wardell arrived in Sydney, he justified his commitment to a new contract by telling the story of old Seville Cathedral which, in 1401, was on the point of falling to the ground. He reminded his audience that the unanimous decision of the clergy then was to 'build the church that those who come after us may take us to have been mad.' When the laughter had died away, Vaughan described how when we go into that most magnificent of all Gothic Cathedrals, we do not think those priests were mad, 'We look upon them as those who come after us may perhaps look upon us if we are true to our resolves, with love and admiration. And, as men in these days, on entering Seville Cathedral, are compelled to think of, and admire, and pray for those whose courage conceived it, and whose love and perseverance built it, so too, if we are equally generous and courageous, if we do our work earnestly and manfully, will those who follow us remember us, and think kindly of us ...'[9]

Vaughan was not merely a spokesman gracing public meetings; he threw himself wholeheartedly into the practical work of raising money for the Cathedral. Over the next two or three years according to his own careful estimation, he personally hand-wrote 1466 begging letters, many of them decorated with amusing cartoons and sketches; and then he would acknowledge each donation great or small.[10] There is no doubt that Vaughan was the engine which drove the building forward, and the engineer who supervised the practical work from day to day was Wardell.

Vaughan was the ideal patron for Wardell who shared the Archbishop's aspirations. They became firm friends and trusted each other's judgement although on some occasions the Archbishop protests at what he judges to be Wardell's extravagance. Where there was disagreement Wardell was respectfully persistent and the Archbishop generally gave way. Here was an all too familiar example of the artist-patron relationship; of the artist who would not compromise his work, who built to the highest standards not only for the present generation but also for the future, and the patron who struggled to find the money to enable him to do so. Wardell's letters to the Archbishop were despatched almost daily keeping him informed of progress, requesting payments, and clarifying points of style and design. In even the smallest matters, Wardell sought the Archbishop's opinion and approval. The correspondence was at times so voluminous that in at least one letter Wardell felt he needed to apologise: 'I am afraid the sight of my handwriting must make your Grace wish I had never learned to write, for I have so constantly to intrude myself upon you ...'[11]

He continues in that letter to argue for properly seasoned wood for the benches. Turning to another subject he objects strongly to a temporary fence erected around the Cathedral works to protect it from intruders. Wardell, throughout his career, was particularly sensitive about decisions taken by others which might reflect poorly on his professional standards and reputation and here he objects strongly to corrugated iron which he believes has been chosen without reference to him. He appeals to the Archbishop to reconsider this decision.

> 'I think it must be admitted that we owe something to the public – the public gave us our ground – and it seems selfish and churlish [to block] the view of a prominent building which should be not only *pro usu civium* but *decori urbis*.[12] There need be no occasion to spend a sixpence more in erecting an open picket fence protected at the top by nails over which the larrikins of the place could not climb with impunity, and behind which they could not hide.'[13]

Like so many other requests, this one was successful and the picket fence was then chosen to Wardell's satisfaction.

If Wardell and his family thought that life and work in Sydney held the prospect of a more placid existence their hopes were quickly dispelled. Wardell thrived on work, and he was soon juggling one commission with another until he had a full order book. He took on a draughtsman, George Denny, and opened an office at Bell's Chambers, 171 Pitt Street, in the heart of the business district.

Before the end of his first year in his new practice he had secured the position of architect to the English, Scottish and Australasian Chartered Bank which, over the following years provided him with a copious and highly remunerative number of commissions for new bank premises across three states. The most remarkable and most admired, is the popularly titled Gothic Bank, now proudly preserved by the ANZ at the corner of Queen and Collins Streets, Melbourne – to be considered in more detail later. Many of Wardell's banks have since been demolished, the saddest loss the handsome, classically-styled ES&A Bank in Adelaide; but others survive – Wagga, Kiama, and Camberwell. At least two banks survive as private residences, one in North Melbourne near the Town Hall, and another, a remarkable house gracing the corner of Mount Alexander Road and Bank Street, Moonee Ponds. A charming feature of the building is the succession of four gables elegantly designed to follow the angle of the turn, under each gable is its oriel window and trefoil moulding. Wardell's fine judgement is evident in the proportions particularly the placing of the elegant sash windows, two of them bay windows. The main construction is in brick with stone facings and the whole affect is romantic Gothic, and most pleasing.

Wardell's remarkable association with the Bank arose from his friendship with Sir George Verdon the Bank's English representative who was appointed Director of Australian operations. Sir George had been a member of Government, holding various Ministries, and for a short time Acting Minister of Public Works while Wardell was head of the Department. He was an educated man, a connoisseur, patron of the arts, and a knowledgeable admirer of Gothic architecture. The two men were fellow members of the Melbourne Club and a friendship developed. As Vaughan was Wardell's partner in ecclesiastical building, so Sir George Verdon was his patron in bank building. Both men were enlightened, sensitive, and appreciative of good design and sound construction. The correspondence between Wardell and Verdon on building projects is as extensive as that between Wardell and Vaughan - a reminder that before telephones all communication, even over short distances, had to be conducted by letter, and only if urgent, by telegraph.

Within three months of settling in Sydney Wardell had expectations of adding a fourth Cathedral to his credit. He had been asked by the Anglican Bishop of Perth, the Right Reverend Henry Parry, to submit designs for the new St George's in its prominent setting in St George's Terrace. Wardell would have remembered the site opposite Government House and the local climatic conditions, from his visit to Perth when he was there to advise the Government on harbour construction four years previously. It is likely that he met Parry at a reception on that occasion.

Wardell sent off a set of plans on August 30, and in a covering letter to the Bishop he wrote: 'I gather from your Lordship's letter that you are anxious to commence the building as soon as possible after you are satisfied with the plans'.[14] He expresses complete agreement with the Bishop that the style should be Early English, 'more fitting than others for the climate and atmosphere of Perth.'[15] There is further correspondence with Bishop Parry after the arrival of the plans in Perth in which Wardell patiently answers queries and confirms that there is no difficulty in applying the alterations his Lordship suggests. 'I have designed the work as you desired it, built in brickwork with stone droppings ... though it would have been infinitely better if it would have been built in stone. My recollection of the Perth bricks is not in their favour.'[16]

Notwithstanding these communications, the bishop and his committee chose Blackett's submission and wrote formally on November 26 accepting his designs. A settlement of Wardell's account for £36.15.0 was made in January 1879. The choice, although disappointing for Wardell admirers, was hardly surprising: Edmund Blackett was Australia's best known and leading Anglican architect in Australia. For the Committee to have rejected his designs for an Anglican Cathedral in favour of his Catholic rival would have seemed provocative in the strained inter-church relations of the times. If Blackett was indeed the favoured architect from the beginning we are entitled to ask why Wardell was approached in the first place.

Wardell had settled his family in a two story Victorian house, Upton Grange, in Edward Street, North Sydney, a municipal district then known as Victoria. Such is the appearance of the house, handsome and well preserved today as part of Shore Grammar School, that we might easily mistake it for one of Wardell's own-designed residential buildings. It was built around 1870 of finely worked sandstone with a steep pitched slate roof, a front gable and veranda. The records show however, that Wardell leased it upon his arrival in Sydney from the owner Francis Hixson, taking over the lease from a Mr W. H. Tullock. At that time Upton Grange was set in spacious grounds with outhouses and stabling for horses. When the Wardells settled there the North Shore was considered a quiet, healthy location, becoming increasingly popular as an alternative to the dirty, crowded city across the water. The population had grown to over 7,000 in the year Wardell moved there, and ten years later, to 19,000. The only inconvenience was the dependence on horse transport and the ferry services between Circular Quay and Milson's point. Long before bridge building, communication with the City was provided by a regular passenger service running from dawn to 7.0 pm daily and charging a threepenny fare.[17] Giving evidence at a 1891 enquiry into the expansion of railway transport one witness stated:

'You must remember that a considerable proportion of the travelling public go backwards and forwards four and five and frequently six times a days. A great many who have come to the city in the morning will go home to lunch, and returning in the evening cross at night.'[18]

The steep gradients leading up from the ferry, and the wet and muddy unsealed roads made walking difficult, although hundreds of office workers were said to have attempted it in the downward direction in the mornings. Horse drawn trams and carriages were the preferred option when returning up the hill in the evenings. The district had to wait until 1895 before the electric cable tram service was inaugurated. The Wardell family was evidently contented with its new Sydney home because it occupied Upton Grange up until and beyond the death of William Wardell twenty years later.

Judging by the short time which elapsed before Wardell was immersed in commissions for the English, Scottish and Australasian Bank he may well have anticipated this employment before he left Melbourne; perhaps Sir George Vernon had already promised his patronage. But there were other banks too. Before the end of the year he was already working on the central city branch of the Bank of New South Wales as well as designs for the ES&A Bank's premises in Adelaide. In the case of Adelaide he sent a request to Verdon for information which shows that the building's setting and environment was as important to him as the building itself.:

'The information I should require would be a plan of the ground with adjacent streets and rights of way; dimensions, levels etc.; existing sewers if any, and levels; photographs of adjoining buildings and their heights; nature of soil for foundations and probably depth from level of footpath. In a city so liable to dust and heat as Adelaide would it not be an advantage that the building should stand back from the street some 10-15 feet to admit of a plantation between the windows and the street – the porch could project to the line of the street.'[19]

The ES&A Bank in Adelaide on King William Street was one of the most imaginative and beautifully proportioned Gothic commercial buildings in the city. It was opened for business in 1880. Sadly we cannot admire it today. To Adelaide's lasting shame it was demolished and a modern high rise bank building erected on the site.

Before the eventful year 1878 had ended, in addition to his bank work, Wardell was sending perspective drawings of improvements and alterations to the design of St Patrick's Cathedral, and submitting his bill for advice and refurbishment of another St Patrick's – a Catholic church in Sydney dating from the 1840s in Grosvenor Street, the heart of the city.

Wardell's correspondence with the tireless Dr Fitzgerald in Melbourne was constant and hardly lessened when Wardell moved to Sydney. 'I shall be only too happy to continue to give you every help I can,' he wrote, 'for I can never lose my intense interest in St Patrick's.'[20] And maintain his interest he did. Until the end of his life he travelled regularly to Melbourne, writing, planning, and overseeing the design and placement of Cathedral furnishings and window designs. So voluminous is the correspondence from Wardell to the two Archbishops on matters relating to each of the Cathedrals that in studying the letters in the archives it is easy to confuse one cathedral with the other.

The year which had commenced with the scandalous dismissals on Black Wednesday, followed by a disruptive domestic trek north, ended with the beginning of a new chapter in Wardell's life. He was now freed from public scrutiny. There would be no more mendacious, libellous attacks in the press, no more government enquiries and ignorant sectarian prejudice. Henceforth, in the role of an authoritative, respected citizen, a leader in his profession, he would run his private practice in the city, taking on more and more secular, commercial work. Apart, however, from continuing his involvement in his beloved St Mary's Cathedral, he would, surprisingly, receive no further commissions for churches in New South Wales.

BUILDING WITH CONVICTION

CHAPTER 19

A PHOENIX FROM THE ASHES

Just as the seeds of the Australian melaleuca and casuarina trees survive the worst of summer bush fires to burst into life when the ashes are cool, so too, although unrecognised at the time, the seeds of Wardell's greatest work, St Mary's Cathedral, germinated in the cooling ashes of one of Sydney's most spectacular conflagrations. To understand how this happened we must go back in time to 1865 just seven years after Wardell arrived in Australia.

On the night of 29 June of that year, after the special evening service of Benediction on the Feast of SS Peter & Paul had concluded, and everyone had gone home, a lone passer-by noticed smoke pouring from the roof of the first St Mary's. He promptly ran to the nearby priory to raise the alarm. The only occupant at that time was Father Garavel who rushed into the burning church to save the Blessed Sacrament and - at great danger to himself – numerous other treasures threatened in the gathering holocaust. But the fire had become so fierce that within thirty minutes the shingle roof and all wooden parts of the building together with precious furnishings were ablaze and beyond rescue. As one eye-witness reported later:

'From the top of the Cathedral clouds of yellow flame and smoke issued, which had a lurid lustre on all around; and at times so bright was the glare that the minute objects in the remotest parts of Hyde Park could be seen almost as distinctly as by daylight, and the reflection in the sky must have been visible for miles around. The interior of the Cathedral was soon a vast furnace of fire, which glowed with intense heat; and the wind and flame roared through the sacred pile, and the timbers crashing above made a noise which somewhat resembled waves beating along the sea shore.'[1]

211

Sydney at that time was hardly a stranger to destructive fires, having no very accessible water system for fighting them. In this case two Volunteer Fire Brigades were quickly on hand and a vast, willing crowd turned up to help, but because the only source of water was found to be a quarter of a mile away nothing much could be done. A hopeless attempt to mitigate the destruction consisted of passing buckets of water from hand to hand. Within a short space of time St Mary's Cathedral became a burnt-out shell. The reaction of witnesses to this awesome spectacle - both Catholics and others in the crowd - was one of dismay and incredulity. The solid Gothic structure, imposing by reason of its height and mass, had become part of the Sydney landscape, a comforting, familiar sight and the pride of the Catholic population. The loss of the building and the treasures inside in money terms alone, amounted to 'considerably over £50,000,'[2] a vast sum in those days.

Archbishop Polding was away from Sydney at the time of the fire making one of his many pastoral visits, on this occasion to Bathurst in the remote Western Districts of his diocese. When word was brought to him it was feared the shock would be too great for the aged man to bear; he was said to be 'prostrated with grief'.[3] Another report made light of his reaction, saying he was greatly comforted by the sympathy of people of all denominations and the determination of Catholics to raise money for the building of a second St Mary's, 'of far more magnificent proportions [which] would speedily be raised upon the ashes of the first.'[4]

The loss of the first St Mary's was so deeply felt that it served to invigorate a popular determination to build anew. Perhaps there was also, among many people, a largely unspoken realisation that the much-loved old Cathedral, now a burnt out shell, was never destined to be the architectural wonder and triumphant mother church of Australia that they believed the rapidly growing Catholic population deserved. Certainly old, idealised sketches of it and one surviving photograph show an unsatisfactory juxtaposition of different Gothic elements, so that the austere south transept (the only one then completed) seemed not to match the more elaborate east end, as though the Cathedral were designed piecemeal - which indeed it had been. Additions to the western facade and the stand-alone belfry were to designs of Pugin; the earlier sections were based on pioneer priest Father John Therry's sketches. The time seemed ripe for a fresh start and Polding recognised this mood and threw himself into harnessing support and raising money.

With this intention a grand public meeting was held at the Prince of Wales Theatre on Thursday July 6, scarcely one week after the fire. The State Governor, Sir John Young, was present on stage, as were the Speaker of the Legislative Assembly and the President of the Council along with leading church dignitaries. Sympathy and support was not confined to Catholics alone. The Archbishop was in the chair and limited his own contribution to a short speech of introduction; it was left to the prominent lay speakers to express the determination of the audience to turn from a tragedy to a plan for reconstruction. His Excellency the Governor explained his attendance by saying he wished to show his respect for the Archbishop and the 'blameless yet energetic manner in which [he] has discharged his duties' and, to prolonged cheers, 'to show sympathy towards the Catholics of this colony, who I believe to be as faithful, as intelligent, and as industrious a class in the community as any that exists.' (Prolonged cheers and applause).[5]

The Honourable T. A. Murray, President of the Legislative Council proposed the main resolution that 'immediate measures be taken to raise funds for reconstruction [of the Cathedral]'[6] and then, to the cheers of the large audience, echoed Polding's hopes that 'the new building would rise far greater in its noble proportions and architectural beauty than that now in ruins.'[7]

This resolution and others were 'unanimously passed'.[8]

Close to £6000 was subscribed that evening of which the Archbishop gave £500 and His Excellency, a member of the Church of England, contributed a handsome £40. Similar fund-raising meetings were held in different parts of the diocese and groups were formed to raise money in the various parishes.

It is not clear how soon it was after the fire that Polding decided to ask Wardell to design the new Cathedral, or whether he took advice on the matter. He would have been influenced by Wardell's work on St John's College and his meeting with the architect at the laying of the Foundation Stone ceremony in December 1859. He may also have met him previously in England and known of his work there. Evidently Polding was not discouraged by Wardell's dispute with the College Council and his subsequent resignation; he may even have sympathised with him. The Archbishop would also have heard first-hand, glowing reports of Wardell's work on Melbourne's St Patrick's Cathedral whose walls and Western facade were sufficiently advanced by late 1865 to promise a remarkable building of exceptional size and grandeur. Clearly Polding would have determined that Sydney should not end up with something of lesser importance. And by choosing Wardell as the architect, a prominent Catholic in the profession with an enviable record for building churches, he would have made a logical choice for the designer of the new St Mary's. The only practical difficulty, Wardell's residency in Melbourne, would not deter the Archbishop.

Polding's first extant letter to Wardell, commissioning him to design a new St Mary's, is dated 10 October, 1865. This indicates that there had been earlier correspondence on the subject, now sadly lost. Polding sets out arrangements for payment of commission and agrees with Wardell's earlier suggestions on such policy matters as the ownership of the plans should the architect die before the work was completed. Polding comments on Wardell's suggestions: 'Nothing could be more fair or liberal on your part and I should be well content to let your arrangement stand.'[9]

Polding then turned to the matter of design and to the much-quoted passage which allowed the architect absolute freedom of expression:

> 'I have but little to say beyond this, that I go to the architect of St John's College to ask him for something that shall again be an honour to himself and to the Catholics of the diocese. The contoured plan which accompanies this will give all necessary detailed information about the grounds. I place everything at your disposal ... I leave it all to you and your own inspiration in the matter and I will not even say that your conceptions shall be restricted to the Gothic style of any particular period. Any plan, any style, anything that is beautiful and grand [is acceptable] to the extent of our power.'[10]

We may pause to wonder at Wardell's work-load at this point in his career, seemingly impossible to comprehend in its magnitude. From this time forward he had on his hands two major Cathedrals in the course of construction, as well as other private commissions; all this in addition to fulfilling his full-time duties as Inspector General of Public Works. The Government Gazettes for those years immediately prior to his acceptance of Polding's commission, confirm what seems to us an intolerable number of advisory appointments. In addition to those mentioned we find him appointed Trustee for the planning and establishment of the Zoological Gardens (March 1862); Member of the Board of Land and Works (July 1862); Member of the Board of Examiners for the Civil Service, Victoria (August 1862); Member of the Commission to enquire into Yarra Bend Lunatic Asylum (September 1862); Member of the Board of Examiners for the Department of Land and Surveys (August 1862); Member of the Commission to enquire into Fine Arts in Victoria (October 1863); Member of the Board to advise on Victoria's contribution to the Dublin International Exhibition (January 1865).

In spite of this work load, he managed to maintain his devotion to his home and family. His fourth daughter, Constance, was born in September 1864, and his third son, Herbert Edward, was born a year later. He was said to be an affectionate father, although in character with the times, a strict one. He was profoundly religious.[11] He had a great devotion to the Blessed Virgin and wrote a prayer consecrating his life and work to Her. He was known as Willy to his intimates, but clearly was not addressed so by his children.

Wardell had turned forty-five at the time of Polding's commission. He was at the fullness of his powers as an architect and no longer was his work to be confused with Pugin's. Although still faithful to Pugin principles, Wardell's work by this time had developed its own character, and been adapted for Australian conditions and requirements. Armed with his professional experience and his religious convictions he was ideally placed to take on the most demanding and most prestigious commission of his career. It was a commission that no architect as ambitious and devoted to the Church as Wardell could refuse. And so, notwithstanding the presence in his office files of the Commissioner for Public Works' admonition to the 'Gentlemen' of the Department that, 'the hours fixed for business of the Department are to be strictly and faithfully conserved for that alone,' Wardell accepted Polding's commission. He replied on October 20 promising that the first instalment of plans would be forwarded 'as soon as possible'.[12]

In the same letter he warned the Archbishop that the estimated cost of between £100,000 and £120,000 was to be taken as a 'general guide' and added that he was sure the church would cost a lot more than that. 'I have already given considerable thought to the question of adopting the style of 13ᵗʰ or 14ᵗʰ century. I believe these to be the best suitable for the climate, and in all other respects desirable. You must be quite prepared for a comparatively costly work owing to the fact that the foundations will be of a most unusual depth.'[13]

Wardell's early sketches of his second cathedral - the elevations and the dominant central tower which show the intended steeples on the south-facing main entrance - differ little from the completed building which we know today. Like its sister cathedral in Melbourne, St Mary's is based on English Decorated principles, but we sense immediately that they are very different buildings; they have two very distinct and different characters. This is not confined merely to the colours of the external local stone in each case – the sombre dark grey of the Melbourne blue-stone and the mid tan of Sydney sand-stone which glows in the afternoon light. The design of St Patrick's with its chevet chapels radiating from the apse clearly owes much to French Gothic inspiration – a similar arrangement can be seen in Amiens and Chartres Cathedrals and the church of S. Ouen, in Rouen, for example. Much has been written about the exterior and ground-plan of St Mary's also being influenced by French Gothic. The West Front (which actually faces south) is said to bear more than a fleeting resemblance to the West Front of Notre Dame. If so, this resemblance is more stylistic than architecturally sustainable. A closer comparison may be made between the north gable with its great eight-light window depicting the Coronation of Our Lady in Heaven, with the east gable of Lincoln Cathedral. Here the similarity in their features is remarkable and a quick glance at the one can lead us to believe – if only momentarily – that we are looking at the other. It is difficult to believe that Wardell did not have Lincoln in mind when designing this part of St Mary's.

For this writer the exterior of St Mary's, particularly when viewing the massive Western elevation from across Hyde Park, is reminiscent of an English abbey church. Like some massive crouching animal or sphinx, it lies in command, unassailable in its natural, elevated domain.

From the moment Polding was consecrated Archbishop in Rome in 1841, he intended that his new diocese should be a Benedictine foundation.

To further this end he had obtained a rescript recognising the old St Mary's as a Monastic Cathedral. For this purpose he had brought Benedictine priests with him to Sydney and he introduced the chanting of the Divine Office at the Canonical Hours. Although the Benedictine plan became impractical over the ensuing years and was resisted by the largely Irish Catholic population, Polding himself was ever reluctant to abandon his dream. Was it not likely, therefore, that when Wardell was asked to design the new St Mary's he should reflect some of this aspiration in the eventual design of the church? There is a distinct character in the exterior of St Mary's reminiscent of the massive English Cathedrals of York, Canterbury, Wells and Westminster Abbey, and even Ripon with its unadorned west towers. Although their ground plans may differ to a greater or lesser extent from St Mary's, and only Westminster Abbey had been a Benedictine foundation, the external elevation and massing of St Mary's has the character of the ancient abbeys with their cloisters and cells for the monks or their chapters of canons. When planning the original St Mary's Polding had wanted a similar arrangement for his Benedictine priests.[14]

One major difference in making a comparison of their appearances is that none of the older cathedrals have spires on their West Fronts. Few English cathedrals do. Salisbury, Norwich and Chichester have central spires, but only Lichfield has spires on its West Front. The problem for architects when adding spires to rectangular towers lies in their need to resolve the horizontal and vertical elements so that the spire does not appear to have been added as an afterthought but presents to the eye an organic fusion of tower and steeple. To what extent this has been achieved with the recently added spires to St Mary's is at least debatable. The intended spires which were part of Wardell's original design and clearly shown in his drawings, appear taller and more successfully fused with the towers than do today's steeples even though erected faithful to his specifications. One is led to wonder whether Wardell, had he lived to complete the building, might have made some modification to his original concept. In Melbourne we know that the spires of St Patrick's were added to coincide with the Melbourne city centenary celebrations in 1939, and that Archbishop Mannix decided that the height, specified by Wardell, should be exceeded by an extra 90 feet and crowned with a Celtic cross. This necessitated reinforcing and heightening the towers to carry the additional weight. On that occasion Wardell's son was invited to undertake the work but loyally refused to interfere with his father's plans. Inevitably cathedrals are rarely completed according to the intention of the original designer. Tastes and requirements change over the centuries and doubtless changes will continue to be made to Wardell's masterpieces in the future. It is curious, and to be regretted, that the same strict regulations in force to prevent clumsy alterations to historic secular buildings, do not seem to apply in the same way to historic churches.

The interior of St Mary's has an altogether different atmosphere from that of St Patrick's. In contrast it is darker, more massive in construction. It is without the lightness, the French delicacy, and youthfulness of St Patrick's, but gives instead an impression of solidity, of greater massiveness, of awe, and maturity befitting the mother-church of Australia. It is as though St Mary's, in keeping with its title, should be the parent and inspiration of the other, yet they are roughly contemporary: St Patrick's conceived seven years previously and consecrated free of debt in 1897; St Mary's consecrated similarly free of debt in 1905.

The differences in the dimensions of the two Cathedrals are minimal. The length of St Mary's is 350-feet while the length along the nave and sanctuary of St Patrick's, excluding the apsidal chapels, is 340-feet. The width across nave and aisles of St Mary's is 80-feet while that of St Patrick's is 2-feet longer. It is the massiveness of St Mary's and its dignity that overawes us. Buildings of similar size today employ an engineer as important as the architect – in the case of the Opera House the name Ove Arup is as celebrated as the architect, Jørn Utzon. Wardell was his own engineer; only an architect who was also an engineer could have managed to balance so successfully the opposing forces inherent in the construction of so massive a building as St Mary's which exudes such confidence and majesty.

Wardell chose 14th century English Decorated style with minimal internal decoration, achieving an impressive unity throughout. Here is an historically accurate model. No one part conflicts in style with another. And yet the whole remains uniquely Wardell. On a dull or rainy morning the prospect in the unlit nave is sombre and dark. But more often, when the sun catches the amber- glass clerestory windows and illuminates the bays, arches and triforium of the opposite wall, and the aisle stained glass windows glow with rich colour, the scene is one of great splendour and visual delight. There is a greater amount of stained glass here than in St Patrick's. The lower light of each window in the groin-vaulted aisles, designed and emplaced after Wardell's death, includes the depiction of important historical occasions in the progress of the Cathedral. Perhaps the most often reproduced is the scene of a convict priest, Father James Dixon, celebrating an illegal Mass in a convict's home. In a scene depicting Archbishop Moran receiving the Cardinal's red hat in 1885, the upright figure of white-haired and bearded Wardell among a group of representative laymen is clearly recognisable. The great north window, the rose window in the south wall, and the east and west oriel windows in the transepts, were first sketched out by Wardell in consultation with Archbishop Vaughan, and manufactured in John Hardman's studios in Birmingham. Wardell was, as usual, rigorous in attending to every detail and demanded the highest standard of workmanship. According to John Donovan, Secretary of the Building Committee at the time, the window in the west transept when received from Hardman's studios and erected in its appointed position, 'had been much admired by competent judges.'[15] But not by Wardell. In another letter Donovan wrote: 'Mr Wardell, after seeing the window which had been fixed in place, has asked me to request you to adopt a deeper tint in the glass of the windows in the future, because of the strong glare of the sun on our windows...'.[16] The request that Hardman incorporate deeper tints for Sydney than would be thought necessary in northern climates, arose from an on-going misunderstanding: the centuries-old habits of northern glass makers were hard to change. Donovan writes again after receipt of the St Joseph Window:

'I am sorry to have to repeat that we are disappointed ... It is generally remarked that it has not the depth and richness of colouring that is present in the Sacred Heart Window. It is obvious that if this blue colour is to be adopted to any extent in future windows, a still deeper toned glass must be used; and no doubt it will sufficiently withstand the glare of the sun. For the Great Northern Window it will be absolutely necessary to use a deeper toned glass than any you have yet used.'[17]

The struggle for deeper tones continued and when the Great North Window was in place in 1885 further dissatisfaction with the light colouring of the glass was noted. It was not until 1886, two years after the first complaint, that Donovan was able to write: 'We are happy to be able to say that this window [the Descent of the Holy Ghost] has given unqualified satisfaction. In richness and harmony of colour and tone it is considered by many persons the most successful of your efforts.'[18]

The minimal decoration within St Mary's - no rich mural painting like the chancel of St Patrick's for example - far from overwhelming the senses, allows us better to appreciate the great gallery of stained-glass on all sides. The near-unadorned sandstone masonry enriched only by the vertical line of the clustered piers with their central shafts extending until they meet the hammer-beam roof springing, leads the eye continually in an upward direction. What adornment Wardell has permitted is confined to the stained-glass, the carved stone bosses at the rib crossings in the aisle vaults, and the heads of the saints carved in stone where the uppermost hood mould of each of the arches terminates above the corbel. Further decorative features can be noted in the quatrefoil tracery of both the six-light clerestory windows and the triforium openings below.

As in St Patrick's the sanctuary has been enlarged and extended forward into the crossing which has the affect of lessening the sense of space at this point. It has been necessary to move the pulpit forward also. The sanctuary's richly ornate brass railings with their quatrefoil tracery designed by Wardell have been retained and when extra lengths were required for the extensions, happily they were reproduced to Wardell's meticulous plans. Also to be found still gracing the sanctuary are the two magnificent lamp standards of burnished brass designed by Wardell. The circular base is supported by lions' feet; the top is shaped to resemble open lilies.[19]

Further delight is revealed behind the High Altar in the ambulatory where the full beauty of the Great North Window depicting the Coronation of Our Lady in Heaven and associated incidents, can be seen in detail. Under this north gable are three chapels dedicated in turn to St Joseph, the Blessed Sacrament (in Wardell's time dedicated to the Virgin); and St Peter. So rich is it in stained glass on three sides that when the sun shines through, the effect is particularly thrilling.

Those who are fortunate enough to know both St Mary's and St Patrick's take pleasure comparing the two, and arguments over their relative merits often arise. Generally it seems St Patrick's has received greater praise

– Robin Boyd's enthusiasm for it has already been quoted. J. M. Freeland a leading authority who was the first to publish a history of Australian architecture, prefers St Patrick's, and dismisses St Mary's as 'tending to squatness, heaviness and dullness, [and] an unsuccessful challenge to St Patrick's'.[20] This strikes us as a hasty and somewhat inaccurate judgement, perhaps illustrative of the patriotic rivalry that traditionally divides Sydney and Melbourne residents - Freeland grew up in Melbourne and practised and taught architecture there for most of his career. Ultimately, the claim to superiority of one Cathedral over the other is misleading and unnecessary. St Mary's and St Patrick's Cathedrals are two magnificent but differing examples of Gothic Revival architecture, two sides of the same Gothic coin. They are supporting sisters, not warring rivals, 'notable for the purity of expression and richness of symbolism [they] rank among the greatest buildings constructed anywhere in that style'.[21]

BUILDING WITH CONVICTION

CHAPTER 20

ARCHITECT OF COMMERCE
AND INDUSTRY

St Mary's Cathedral, the crowning achievement of Wardell's Sydney years, so dominates our common knowledge of him that his admirers are apt to pay little attention to his other, lesser-known works. And yet his commercial work of this period, his banks which survive, and his warehouses, are visually striking, never dull, and the best of them rate as minor masterpieces of their genre.

The puzzle is that, once established in Sydney, he undertook no further church building throughout the colony, other than supervising progress on his Cathedral. The diocese was larger in area than it is today, new parishes were springing up so that Archbishop Vaughan is credited with doubling the number of churches and chapels in New South Wales within the space of ten years.[1] Given Wardell's unrivalled reputation for Catholic ecclesiastical work in Victoria, and the warm relationship which was soon to develop with Vaughan, we are left to wonder why he ceased designing churches once he left Melbourne. Reed, his chief rival, was left behind in Melbourne, and Blackett, who had until then dominated church building in Sydney, was to die in 1883 leaving an architect of Wardell's capabilities free to take his place. This, however, did not happen.

Wardell's work in Sydney, apart from the Cathedral, was curiously subdued and to the general public largely anonymous. The high expectations forecast by the *Sydney Morning Herald* following his arrival in the colony, seemed not to materialise. Wardell, the architect of the grandest Government House in the entire continent, and formerly the Head of the Department responsible for some of Melbourne's finest public buildings, would create no public, secular buildings for Sydney during his remaining twenty years of residence there. Part of the explanation may lie in the economic realities dividing the two colonies. As Freeland notes,

Melbourne was richer than Sydney, and Melbourne being younger had a greater need for new public buildings. Backed by the Victorian gold discoveries she could afford to indulge herself. By the time Wardell arrived in Sydney, the city was a hundred years old and 'was already supplied with a number of adequate if not particularly desirable buildings.'[2] Any opportunities remaining for new building work were mainly limited to commercial and industrial needs. Wardell, an astute businessman as well as an artist, would have recognised this need and profited by it.

Sadly, the most substantial Wardell building, the Union Bank which would have marked his presence in the heart of Sydney's business district and perpetuated his fame there, was demolished and replaced by a modern glass office block in the rush to modernise the city in the 1960s. Here, on the sharp corner where Pitt and Hunter Streets meet, was a handsome, four-story classical building crowned with a dominant belvedere tower, the style bearing more than a fleeting resemblance to Melbourne's Government House. Now, all that is left to remember it by are photographs, one showing the stately banking hall with a cross-vaulted, coffered ceiling, and two rows of pillars resting on marble pedestals.[3] The exterior shows an unusual-angled elevation made necessary by following the sharp bend of the corner. The main entrance had a classical portico graced with pillars and pediment. Also pedimented in the classical style were the second floor windows, larger than those of the first floor. The Hunter Street frontage was topped with a balustrade which echoed the roof style of Melbourne's Government House. The stately belvedere was an extension of the slightly projecting corner tower, but gave the impression of being too dominant and top-heavy in comparison with the rest of the building. The demolition was a sad loss to the city and one which leaves the former New South Wales Club close by at 31 Bligh Street as the only example of Wardell's classical work within the city centre. Designed by Wardell for the Club and built in 1884, it is now preserved in the care of the National Trust but sublet as exclusive company offices. There is something remarkably brave and defiant in its will to survive. Hemmed in as it is by monsters, faceless, alien and threatening, one fears for its future. The simple proportions, dignity and grace are an uncomfortable reminder of what we have lost in the rapacious city landscapes today. Wardell seems to have had Charles Barry's Reform Club in London in mind because the exterior of 31 Bligh Street is remarkably similar to it. Originally of only three stories, the style is what might be described as 'Italianate Classical,' the interior larger than the frontage suggests. The house is built around a central courtyard allowing the handsome rooms to open onto a pleasant garden. The main hall has a beautiful iron balustraded stairway and a painted ceiling of red ochre and gold.

Another important city building by Wardell to fall victim to the bulldozers was the Union Club, situated directly opposite the New South Wales Club in Bligh Street. The Union had a larger, grander frontage than its sister across the road but was designed in the same Italianate classical style. Had it survived, Bligh Street might have aspired to be Sydney's Pall Mall, the home of London's most prestigious Victorian clubs. The *Sydney Morning Herald* of 5 March, 1887 enthused, 'The facade facing Bligh Street presents as its central feature a massive portico thirty feet in height from the pavement line, forty-two feet in length and nine feet in depth, with six bold columns flanked by well-balanced pilasters; each angle of the portico being finished off with coupled columns at the proper distance apart. The height of the ground floor is eighteen feet, that of the first floor, fifteen feet, and that of the second floor twelve feet…The whole structure is balanced and dignified.'[4] What strikes us in the photograph – the only way we can now judge - is the elegant detail and the grandeur of the portico, the low relief frieze below the cornice, the balustrade again echoing Government House, and the restrained detailing of the facade. Clearly no expense was spared.

Before the Union Club was completed there arose another of those occasions in Wardell's professional life where he thought his honour, or in this case his conduct, had been impugned. The injury, real or imagined, was contained in a letter from the Union Club committee wherein it was claimed the committee had been deprived of the opportunity of conferring with Wardell due to his temporary absence from Sydney. Wardell, ever sensitive, thought the letter contained 'very unjust implications, or to say the least, that such an implication would be fairly deductable from it.'[5] He then explained that the exigencies of his work required him occasionally to absent himself from Sydney, but 'on no occasion have I left Sydney without being adequately represented.'[6]

A close study of the Wardell files of extant professional correspondence reveals many examples of his sensitive personality, a man jealous of his reputation, easily insulted, and meticulous in correcting wrongful accusations. A later example of this highly-tuned sensitivity to perceived imputations, and how strongly Wardell reacted to them, is revealed in a letter addressed to Dean O'Haran, Private Secretary to Sydney's third Archbishop, Cardinal Moran. Wardell complains that a letter from Dean Mahoney questioning Wardell's account, casts a slur on him, 'which may lead to a very wrong estimate of my behaviour in the matter and convey the idea that in my professional dealings with Cathedral matters I have evinced a greedy gain-seeking of which I am totally unconscious.'[7] In a long letter of justification he sets out the history of his association with the Cathedral, the original conditions of his employment commencing with the commission from Archbishop Polding. He relates how in the early stages he could seldom leave his employment in Melbourne, but visited the works in Sydney as often as he was able, 'with a very natural solicitude for the success of the works which from every point of view were so full of the deepest interest to me.'[8] On these occasions he was paid neither a fee nor expenses and it is clear from the correspondence that the Dean of the Cathedral had expected this arrangement to continue gratis. But the circumstances changed.

Wardell goes on to explain that when he came to live in Sydney, at Archbishop Vaughan's request he involved himself much more closely with the works and design matters. 'I visited them daily and sometimes two and three times a day, in short I gave all the assistance [I could] and never thought of being paid for it.' But later when the Archbishop formally engaged Wardell for assistance in designing furnishings and other details he 'asked me again what my professional fees might be'. Wardell responded by setting out his standard fees in a letter but, he adds, 'I had no communication as to fees and therefore in the Cardinal's absence charged the usual 2½ per cent for superintending the works.'[9]

Wardell seems to have had no such misunderstandings with his friend and patron, Sir George Verdon, who was at the time Inspector and General Manager of the ES&A Bank in Australia. To Verdon, generous sums of money spent on designing fine new buildings for the company was money well spent. So much so that that the parent company in England grew alarmed at the amount Verdon was lavishing on the Melbourne head office – some £50,000 – and sent to Australia one of their English directors to report on it. His praise for 'the finest bank building in the colony' satisfied the dissenters so that no further objection was raised.[10]

Fortunately this 'finest bank building in the colony', now the ANZ, has been proudly preserved, and attracts admiring tourists in addition to its many customers in the city.

George Verdon, born of a clerical family in Lancashire and highly educated, came to Australia at age seventeen and although he was well connected, his early business ventures including a visit to the Gold diggings, were unsuccessful. It wasn't until he entered politics that his fortunes changed. He rose to the post of Treasurer in the Victorian Government and supported several popular, worthwhile causes so that he 'acquired the esteem and respect of everyone'.[11] He was a member, along with Wardell, of the 1863 Royal Commission on Fine Arts. He helped to establish a National Museum, the National Gallery of Victoria, and the Royal Mint, and negotiated an advantageous loan from the home Government. Already a member of the Board of the English, Scottish and Australasian Chartered Bank he was knighted in 1872 while on a visit to England. Verdon's interest in the arts was genuine. He had travelled widely and was an expert in his favourite Gothic period of architecture. A Gothic tombstone designed by Wardell (as also was his coffin), adorns his grave in the Kew cemetery. It was Verdon's decision as Director in 1883, that the ES&A Bank headquarters in Melbourne should be Gothic in style, and that his old and admired friend Wardell, although then living in Sydney, was the man to design it. A voluminous correspondence, opinionated and occasionally argumentative, flowed between them, and survives to show the extent of the detailed planning and concern for detail that occupied them both for close on three years.

The Bank's appearance on the corner of Queen and Collins Streets is remarkably similar in detail to the Venetian Palazzo Ca D'Oro and the Doge's Palace. But what strikes us as we pay it more attention from a vantage point across the intersection, is its symmetry and proportion; although ornate when compared to the modern, utilitarian office blocks close by, it is not overly so; the Gothic features are restrained and tastefully accurate in contrast to the later, over-embellished wedding cake frontages either side of it. Unlike these, Wardell's frontage is integrated in the whole concept, and the style is carried through to every part of the interior including the Director's residence on the top floor. As the *Illustrated Australian News* noted approvingly at the time, 'We are glad to see in Melbourne, in a purely commercial building, the old rule of truth in construction and honesty in material, being regarded.'[12]

The first-time visitor upon entering the main door in Collins Street marvels at the beauty of the richly decorated banking chamber with good reason. It is so unlike any banking experience elsewhere. After the initial surprise at the palatial vista, a closer study of the decoration and the detailing shows how meticulous both Wardell and Verdon were in engaging only the best craftsmen, and using the finest materials irrespective of expense. All woodwork and furniture were solid Tasmanian Blackwood made by a master craftsman named Rose; the interior painting and decoration was carried out by an expert named Wells who was brought from Scotland especially for this purpose. The ceiling is hand-painted, and in order to achieve the decorative patterns thousands of sheets of Gold Leaf were used. There is a touch of Pugin's influence here, re-emerging long after Wardell had established his own distinctive style. Verdon, whose contribution to the overall appearance of the bank should not be underestimated, was also an admirer of Pugin. His suggestions were not always appreciated by the architect.

The 'Banking Chamber' is dominated by two rows of cast iron pillars supporting plated rolled-steel joists reminiscent of St Pancras Railway Station or, closer to home, the Library of St John's College; the capitals of each pillar are decorated with wrought-copper flowers and foliage. The whole aspect inside is one of sumptuous detail, and is so contrary to our experience of banks that we can only stand and admire. The outspoken critic and architect, Robin Boyd, judged the Gothic Bank as 'probably the most distinguished building of the whole Australian Gothic Revival era, not forgetting the cathedrals.'[13] Evidently the Melbourne public agreed; a poll conducted by the *Age* newspaper resulted in 'The Gothic Bank' being voted the most popular commercial building in Melbourne.[14]

The first decade of Wardell's residence in Sydney was mainly a period of consolidation. He established a sound practice under the name Wardell and Vernon, supported in the main by bank and warehouse designing, and the occasional consultancy; underscored, like counterpoint, by his continuing devotion to work on St Mary's Cathedral. The Church continued to be an important part of his life, both in a deeply personal sense and, in spite of an absence of major commissions, professionally important to him also.

In 1880 Father Joseph Dalton left St Ignatius Church in Richmond and moved to Sydney to established St Ignatius's College famously known now as Riverview, after the district. Wardell's youngest son, Herbert Edward, aged fifteen, was enrolled as a foundation day pupil. In a photograph of Upton Grange we see the figure of Herbert's vigilant, slightly portly but evidently proud father, encouraging his son as he mounts his horse before riding off to school - riding a horse, or rowing across the harbour were popular means of reaching remote Riverview in those days. Wardell was appointed Consulting Architect to the infant College and was responsible for designing the first substantial building there, known as St Michael's House. It is still in use today, a handsome two-story Victorian Gothic building with a steep pitched roof of slate, end gables, and a dominant projecting tower-like entrance having its own gable at roof level. The satisfying proportions and handsome chimneys on the roof line are distinguishing features and must have inspired confidence in the parents of the first pupils. The opening on 20 September, 1880 was marked by a grand dinner at which Wardell was present and at which he was publically thanked for his work.[15] But his relationship with the Building Consultative Committee deteriorated when he strongly criticised the poor quality of the bricks used to construct St Michael's House, and urged them to change their supplier. The relationship deteriorated further over the plans for the Infirmary, also designed by Wardell but not carried through.[16] A note in Father Dalton's diary of 17 September, 1883 records tersely that he 'gave Mr Wardell notice as the Consulters did not approve of his plans. We will employ another architect.'[17] Whether Dalton, an old friend of Wardell's, approved of the dismissal is not revealed. On this occasion, as on so many others, Wardell was right in his professional judgement because the bricks used in building St Michael's soon started to crumble away. Today the house is coated with a grey cement rendering.

Wardell had a happier relationship with the Marist Brothers when he designed a new school for them on the North Shore.

Passengers on ferries leaving or arriving at Circular Quay may catch a glimpse of a castellated square tower and a second tower capped with a romantic Gothic steeple just visible above and beyond the functional 1960s steel and glass Passenger Terminal on the western quay. Unknown to most of those passengers the architect of this one-time headquarters of the Australasian Steam and Navigation Company was also responsible for St Mary's Cathedral. Apart from sharing the same architect the two buildings have much in common. The A.S.N. Company building is one of the most distinguished and strangely enchanting commercial buildings in Sydney, unsurpassed by the other historic buildings in the famous Rocks area. One side faced Sydney Harbour (until the Passenger Terminal was built) and the opposite side on Hickson Road comprises a succession of four gabled sections capped by stepped Dutch gables. The construction is colonial brick throughout and the visual interest rests mainly in the variety of aspects and romantic shapes. Here is a rare example of a purely commercial building that is both functional and of lasting visual interest and fascination; of an era when aesthetics was an important constituent of commercial prestige.

The second of Wardell's three monumental warehouses in the area lies at the other end of Hickson Road on the waterfront where it doubles back round on the western side of the Bradfield Highway. A massive brick elevation by the standards of the time, it is graced by similar stepped gables. Because it was a bond warehouse where the prime need was for security of imported goods, the impression from Hickson Road is somewhat severe. But the elevation is relieved by the use of deeper-toned bricks in continuous bands and in the capping of the windows. These details, together with the handsome rounding of the corner elevation make the whole affect a visual delight. Wardell's genius is best demonstrated by its elegant proportions, and a pleasing symmetry. In recent times the vast interiors of the warehouses have been tastefully restored and developed to incorporate art and craft studios, shops and offices.

Before we leave Wardell's secular work and return to St Mary's Cathedral, mention must be made of his consultancy work for the Auckland Harbour Board concerning their proposed Calliope Graving Dock. His reputation as an authority on harbour engineering, confirmed by his work in Victoria on the Williamstown Dock and Warrnambool harbour - if not by his work in Western Australia - was evidently recognised in the neighbouring colony of New Zealand. He had been requested to examine Plans and Specifications prepared by a Mr McDonald, and his report to the Board is dated 27 July, 1883. His comments include a suggestion that the total length of the Dock be divided into two divisions so that 'it has the advantage of admitting a greater economy in working, affording facility for accommodating two ships at once, and for allowing the outside ship to be removed without disturbing the other.'[18] He recommends the use of Portland Cement below the high water line and ends by congratulating the Board, and praising the designs of the Engineer, Mr McDonald.

It is ironical that at about the time Wardell was writing his report on the Auckland Dock, one of the most brilliant harbour engineers of the period, Charles Yelverton O'Connor, was in the employ of the New Zealand Government. He had only recently designed improvements to the Hokitika and Greymouth harbours on the dangerous west coast, and seven years later would be lured to Western Australia where he would successfully over-turn Wardell's proposals for Fremantle Harbour.[19]

CHAPTER 21
A POSITION OF EMINENCE

However much attention Wardell gave to his many commercial and secular works during those early years of residence in Sydney, and important as they obviously were to him, it is clear from his voluminous business correspondence that his over-riding interest and concern was for his two Cathedrals: and since it was on his doorstep and in greater need of attention, St Mary's Cathedral in particular. He admitted that he had for the building, 'a very natural solicitude for the success of the works which from every point of view were so full of the deepest interest for me.'[1]

In 1883 Wardell celebrated his sixtieth birthday, a time in life when most professional men may justifiably think of retirement. Retirement for Wardell the artist and engineer, however, seemed never to be considered. His health, which had given him so much concern as a younger man, held firm so that he continued visiting his office, his Cathedral, and his Club in the city until two months before his death. He became, in those later Sydney years, a leader and authority figure in his profession. In his sixtieth year he was elected Honorary Secretary of the newly formed Australian branch of the Royal Institute of British Architects (RIBA). He was elected a member of the Royal Society of New South Wales in the same year, and founded the New South Wales Branch of the Royal Geographical Society of Australia. Wardell had become a man of importance in the community. His receding white hair and white beard, his intense blue eyes and his upright posture added gravitas to his appearance. As we study the last photograph we have of him seated in his library, it is easy to imagine that even the high and the important in the land would have deferred to someone in possession of such wisdom and experience. Certainly there is much evidence from their correspondence that Cardinal Vaughan in Sydney, and Archbishop Carr in Melbourne regularly did so.

In July 1881 Wardell put forward a plan at the request of the Archbishop to enable an incomplete section of the church to be opened and used for services two years later. Wardell explained the work that was necessary to achieve this aim. As well as pressing on with the new building contract costing £21,000, he added a list of the furniture that should be ordered without delay along with estimated costs: the candlesticks, lectern, Tabernacle, flower vases and many other items. He ends his letter by warning Vaughan, 'I don't know whether anything has been proposed for the font, but I may state for your guidance that I do not think anything suitable for the Cathedral could be had for less than £100.'[2] In another letter he urges James Bell, his trusted cabinet-maker of Richmond, Victoria, who was making the benches and footstools, not to be hurried over the work. 'Everything must be done in the very best manner.'[3] In August he writes to Pugin's former partner and craftsman, John Hardman of Birmingham, to commission a design and make candlesticks, a crucifix, and Tabernacle. 'If they leave England by an Orient steamer by or any time before the end of the current year [1881] they will arrive in time [for the Opening]'.[4] When Vaughan and his Building Committee argued for some cost-cutting measure as they did when hard pressed occasionally, Wardell would counter their proposals politely but firmly. When the Committee was determined to install wooden ceilings for the aisles in place of Wardell's proposed stone groin vaults, he presented this unanswerable argument in response:

> 'I may be pardoned for urging that in carrying out a vast work like this –
> built not for today but for all time – the supreme consideration is not what is
> cheaper, but what is best, and if one method is better than another I venture to
> think it should be adopted although it may take a longer time to complete.'[5]

As we look up at those aisle vaults today with their decorative bosses we rejoice at his persistence.

One of Vaughan's decisions which distressed Wardell but which he had to accept, albeit reluctantly, was the staging of the money-raising *Ye Fayre of Ye Olden Time* within the walls of the unfinished Cathedral nave. Open to the public for a whole month following Easter 1882, only five months before the official opening date of the Cathedral, the interior was turned into what was imagined to be a sixteenth century market place with stallholders in Elizabethan dress, Punch and Judy, Fortune-tellers, minstrels, maypole dancers and the provision of various competitions and games and other attractions. A contemporary illustration of the scene - clearly idealised for promotional purposes - shows a remarkable but incongruous transformation of the Cathedral interior, displaying a considerable investment of initiative, money and talent.[6]

One can understand Wardell's disquiet. He was concerned at the possible disruption to his construction work, but more pressing perhaps was his fear that his reputation might be damaged by an erroneous belief that he personally was associated with the project. Here was another instance of his acute sensitivity to any implication of poor taste on his part, or the perceived lowering of his standards. One of his objections was the choice of lights used at the *Fayre*, '... the public will, of course, suppose that I am responsible for them. I must therefore beg Your Grace in justice to me to make it publically understood in that case, that they are put there against my advice and entreaty.'[7] How His Grace was going to make this announcement 'publically' without both Wardell and the Archbishop sounding slightly ridiculous he did not reveal.

Their friendly relationship and respect for one another survived these minor differences and in most instances the Archbishop supported Wardell's requests and took his advice. They both had a sense of humour and when a well-meaning parishioner presented the Cathedral with a statue of St Patrick devoid of artistic merit, neither of them could think what to do with it without giving offence to the donor. Vaughan tossed the problem to Wardell but he was equally uncertain what to do and took refuge in sending a newspaper cutting to His Grace 'for his amusement'. It was a humorous report of a minor court case in which an old timer was accused of a misdemeanour and pleaded his innocence from the dock:

'Your enquiry stumps me the darndest. The more I think on it, the more i cannot tell. As near az i can rekolek now, I think i don't kno. Much mite be sed both ways....'[8]

Evidently the Archbishop and his architect were unable to resolved the problem, or perhaps the Archbishop felt less offended by it than Wardell, because the unwanted statue remains there today.

In spite of the *Fayre,* preparations continued throughout 1882 for the Solemn Opening and Blessing of that part of the Cathedral already completed - the northern section with its dominant traceried window, as yet with plain glass, and the west transept. Throughout the year Wardell plied the Archbishop with reports of progress and requested his agreement to further expenditure. In March, in his letter in which he apologises for the frequency of his requests to the Archbishop[9] – quoted above - he continues with a defence of the benches which he designed and to which, apparently, the Archbishop raised some objection on the grounds of cost. Wardell explained that it was necessary to use wood well seasoned. Later when the benches were delivered a month before the grand opening, the Archbishop informed Wardell of criticism being expressed as to whether they were of sufficient comfort for worshippers during the long ceremonies planned for the Grand Opening. Wardell advised his Grace to allow them to remain as they were, 'It is an absolute improbability to fit them to everybody's back, and it is an easy matter for each person to accommodate his position to them. The seats are wide enough to allow anyone to sit at any angle he pleases ... I venture to say that [they] will not be objected to when people are accustomed to them.'[10] The benches remained as designed by Wardell, and the Cathedral was ready for Opening as he had promised. The new, complete peal of eight bells, a personal gift from Archbishop Vaughan which he had purchased in London in 1881, rang out for the first time on the morning of September 8, 1882. The day, or rather the three days, of celebrations marking the Opening and Solemn Dedication of the Cathedral was held over the 8th, 9th and 10th of September.

> 'The glorious event – so long and so lovingly, although somewhat anxiously, expected by the Catholics of New South Wales is at length an accomplished as well as an historic fact. After a struggle of nearly seventeen years, relieved, it is true, by many bright episodes, the Metropolitan Church of this Province, and the Mother Church of the Australias has been opened for Divine Service.'[11]

A contemporary report of the first day of the Triduum went on to describe how the Cathedral presented an impressive scene:

'Although unfinished, with its clerestory closed against the light of the sun, and its temporary roof bare and unadorned, still the symmetrical proportions of the building, the groupings of the massive pillars, sustaining graceful pointed arches, the fine expanse of the sanctuary with its numerous steps, the shapely altar and throne of the Blessed Sacrament, sculptured out of white Oamaru stone and supported in front on pillarets of Irish marble – formed a pleasing whole – a picture not only gratifying to the eye, but very impressive.'[12]

The many important visitors, Bishops and clerical guests from out of town were accommodated at St John's College where they lunched and dined in the Archbishop's dining-room and were conveyed to St Mary's for the multiple ceremonies in carriages provided. The opening grand procession, to the strains of the organist's especially composed march for organ and orchestra, signalled the start of the ceremonial Episcopal High Mass. Everything had been planned and rehearsed to make this a glittering, memorable occasion – a demonstration of the Church in a triumphant mood.

'The sanctuary at noon presented a very impressive appearance. The full yellow light, streaming in through the great north window gave a rich but toned irradiation which helped to heighten the contrasts presented to all round.'[13]

Among the invited guests seated in reserved places in the Cathedral were Frederick Darley, the Chief Justice; and, in his official robes, the Mayor of Sydney, Sir Matthew Harris; the Town Clerk; and various office holders in the Legislative Assembly and the Council; visiting Mayors and Consuls and leading Public Servants. Although they are not mentioned among these, surely the cathedral architect and his family would have been included among the guests. It would be another forty-six years, long after Wardell had died, before his Cathedral would be completed and opened with a repeat fanfare and an estimated 150,000 persons present on 6 September, 1928. Architects of cathedrals seldom if ever live to see their work completed.

In the two years 1883-85 Wardell was primarily occupied with his work on banks and warehouses already described in the previous chapter. As a concession to his age perhaps, and to ease the work-load for one man, he took on a partner into the practice; from 1 October, 1884 it became known as Wardell and Vernon.

Walter Liberty Vernon had followed much the same career path as Wardell. Twenty years younger than Wardell he was born and educated, and completed his articles in England He went on to have a successful practice in Great George Street, Westminster, but like Wardell, his health suffered from London's climate. Recurring bronchial asthma forced him to emigrate, and he chose Sydney as his destination. Vernon, described as being 'consumed by his interests in architecture and soldiering,'[14] clearly shared much in common with Wardell. His efficiency and his high professional standards in the office allowed Wardell freedom to travel to his various projects in other states while knowing that his work in Sydney was being well supervised in his absence. Wardell made visits to Melbourne, and to Tasmania in May 1885. On the latter occasion the Secretary of the Union Club complained that 'the committee had been deprived of the opportunity of conferring with us.'[15] On his return to the office Wardell was upset by what he believed to be 'very unjust implications ... on no occasion have I left Sydney without being adequately represented.'[16] Although it is not clear from the office correspondence, the reason for his visit to Tasmania was likely to have been in connection with the failed building of St Mary's in Hobart. His letter written three years previously in which he disowns responsibility for the debacle, has already been quoted.

Wardell returned to Sydney to learn that his new Archbishop, Francis Moran, who had arrived in Sydney to joyous Irish acclaim the previous year, had most recently been created a Cardinal on July 27 that year while on a visit to Rome. Archbishop Vaughan had died on 18 August, 1883 while on a visit to his home in England. He was prematurely worn out, a year short of his fiftieth birthday. With the arrival of Moran the Irish faction had succeeded in breaking with the English Benedictine tradition, having lobbied in Ireland and Rome for the appointment of their own man. In several ways Moran was a controversial figure, lacking the diplomatic and negotiating skills of his predecessor. He was forthright and intensely dogmatic in his views. 'Moran's combativeness may have pleased most Catholics, but it was driving the sectarian divide ever deeper while the Sydney archdiocese became ever more Irish'.[17] But he was scholarly, a hard worker and energetic. He is chiefly remembered for his successful advancement of Catholic education and for upholding

Catholic tradition. He was also a builder - St Patrick's College, Manly was one of his triumphs - and pushed ahead with raising money for the completion of the Cathedral. His biographer, Philip Ayres, notes that in assessing the degree of fund-raising which made the completion of the Cathedral possible, the approximate proportion of money raised by the two Archbishops when taken consecutively, may be judged as one third by Vaughan, and two-thirds by Moran.[18] Moran's determination to support the Cathedral building programme and his success in raising money is acknowledged by the larger-than-life statue of the Cardinal erected on one side of the south portal of St Mary's. The twin statue the other side is not of Vaughan as one might have hoped, but of another Irishman, Michael Kelly, the fourth Archbishop of Sydney from 1911 to 1940. Even if the Irish dominance created a divide among people in the Church at that time, it did not affect Wardell's relationship with his new Archbishop. Moran's determination to work for the completion of the Cathedral and the respect he paid to his architect, ensured a harmonious relationship between them.

Wardell's serene life at Upton Grange, North Sydney, was struck with tragedy 1888. It could not have been a happy year for the family. In June Willie's beloved wife, Lucy Anne, first showed signs of a form of dementia. She was cared for by her two unmarried daughters, Ethel and Constance, but after four months in which her condition deteriorated, she died on October 31, and was buried on November 2 in the family grave at Gore Hill. Lucy and 'Willie' had been married forty years and there is much evidence that their union had been a close and happy one. Wardell's letters from Upton Grange from this time forward are edged with a black border.

Wardell's eldest daughter, Mary, then aged forty, had married John Joseph Power, and their son, Joseph Wardell Power, born in the year after his grandmother's death, became a medical practitioner by profession but remained an artist by inclination, inheriting his grandfather's talent. He was later to endow the Faculty of Fine Arts in the University of Sydney. Wardell's youngest son, Herbert Edward, aged twenty-three at the time of his mother's death, followed his father into architecture, and joined him in the practice.

Progress on the Cathedral works continued apace. A contract for the completion of the tower, including the roof and the tower arches, had been let to Mr John Fry who submitted the lowest tender at £21,500, 'To be paid over a period of five years by quarterly instalments...'[18] Late in December the Cathedral Finance Committee raised the question of Wardell's fees in connection with the John Fry contract. Wardell replied on 2 January, 1889 and while showing that, in the case of Cathedral work, he was willing to relax his otherwise strict conformity to architectural practice, he still expected a reasonable payment for his work: 'I am, and have always been perfectly willing to make any reasonable concession, not because of the extent of the work, but because of its character and destination. I left myself entirely in His Eminence's hands as to what my remuneration should be for the altered plans for the Archbishop's House and for the work's on the north gable.'[19] He concludes by suggesting a meeting with the Committee to 'arrive at an equitable arrangement.'

As the decade drew to a close, Wardell's wide experience and recognised probity resulted in his services as a consultant being frequently employed both by professional colleagues and by government agencies. Following the death of one man and the serious injury of two other workers at a quarry near the Prospect Waterworks, it was Wardell who was asked by the Under Secretary of Public Works, J. Barling, to investigate and report on the circumstances. He found that on Saturday 7 December, 1889, three men were drilling the rock face from a position on a shelf fifteen feet above the quarry floor. A large rock weighing about twenty-five tons was displaced above them and fell, smashing through the shelf and crushing the head of one, George Goodford.[20] Wardell made his inspection of the quarry on January 2, 1890, to find out whether the accident was due to neglect by those in charge. His report reads in part: 'In the natural alarm at seeing the rock falling [the men] rushed to escape it, unfortunately in the wrong direction ... The accident appears to us to be one that could not have been foreseen, and we do not think that any charge of neglect in respect of it can be supported.'[21]

Wardell at the age of sixty-seven was as busy as ever in that January of 1890. He was working on designs for a convent school in Kew, E.S.& A. Banks in Surry Hills, Grenfell and Orange, and a mausoleum at Rookwood for the family of a rich patron, the Honourable John Frazer. He was angered when heavy rains flooded the organ chamber of St Mary's and damaged the new organ. He authorised repairs to cost £106 and injudiciously blamed John Fry of negligence 'in not sufficiently protecting the building according to the terms of your contract and which sum will be deducted from the next instalment that may become payable to you.'[22] He has to apologise to Fry when he learns subsequently that Fry had not left the roof of the organ chamber insufficiently protected 'as was first surmised.' He then lays the blame on the 'extraordinary rainfall' and advises that the gutters should be regularly cleaned. In March he writes to the Building Committee asking that 'the present construction design ... not my design...be withdrawn'[23] and his original groined vaults in stone substituted.

Wardell continued to contribute advice on building work on St Patrick's Cathedral where the central tower was still to be erected. In June he asked Archbishop Carr to return his plans to him in Sydney so that he could prepare estimates for that stage of the work. 'I will, of course, bring them back with me when I return to Melbourne.'[24] In the event he sent the plans and specifications back to the Archbishop by R.M.S. '*Ormuz*' leaving Sydney on July 21. The package consisted of the section through the transepts; the section through the Sanctuary Aisles; the Sanctuary Roof, Ceiling and details, and the South West Tower and Gables. All of which suggests that much still remained to be done to St Patrick's before the Official Opening and Consecration on October 31 seven years later.

Throughout 1891 Wardell seemed like a juggler balancing his commitment to one Cathedral in turn with the other, or like a chess master playing a game against two opponents at one and the same time. In May the correspondence dealt mainly with St Patrick's - designing Confessionals, recommending that they should be made of French Walnut, recommending facilities for fire precaution; comparing the costs of different types of marble for the altars; and warning the Archbishop of the risks in allowing Italians to interpret Gothic design: 'I had a bitter experience of that here, in Archbishop Polding's tomb. When it came I hardly recognised my design, for the details were so atrociously rendered, and I felt how seriously unfair it was to me. The Italian carvers of course do excellent things in their own (Italian) style, but not in Gothic style.'[25]

In July he wrote to a Melbourne architect to ask him to inspect storm damage to his Convent School in Kew. One notes how well he remembered his lesson in making too hasty judgements, as he had done over the St Mary's Organ Chamber. In this instance he was more cautious in apportioning blame. He wanted to know 'whether there was any real fault in the work (I should be very sorry to blame the Clerk of Works without cause) or whether the injuries might be expected from the extraordinary weather.'[26] The architect, Todd, reported on the work 'most favourably' but added that the windows in the Lady Chapel may have been defective. Wardell politely asked the contractor to remedy this. He visited Melbourne in October for consultations and inspecting the work and when he returned to Sydney he advised that the great doors of St Mary's should be cleaned and oiled 'I would not recommend varnish at all, and certainly not where the sun has any action.'[27] He ordered Tabernacles for St Patrick's from Hardman in Birmingham and insisted that they should be 'of the highest class of work and richly jewelled [and] fitted with the best procurable lock.'[28]

The year ended as he was considering electricity installations in St Patrick's, and ordering '500 incandescent lights of 15 candle-power each.

At the age of sixty-eight Wardell was as sharp and clear-thinking as ever, opinionated yes; testy perhaps; and certainly showing an old man's increasing impatience with inefficiency and falling standards. He conducted a lengthy, accusatory correspondence with the publishers over a missing issue of his professional journal, the *Builder*.

The heavy commitment continued throughout the last years of his life; Wardell showed no sign of retiring, or putting away his drawing implements, nor of cutting down on his correspondence. He would remain active and productive to the end.

CHAPTER 22

THE FINAL YEARS

Wardell had removed his practice to chambers on the top floor of the Bank of Australasia's new premises at 259 George Street, overlooking the intersection of George and Jamison Streets. With the resignation of George Vernon who joined the Public Service as Government Architect, the firm became known as Wardell and Denny.

The building boasted an electric lift, a welcome facility for the near-septuagenarian architect on his frequent visits to his city office. But those early, clanking, wood-panelled cubicles with the traditional metal-grating doors, were still fairly rare in 1892 Sydney, and lacked the modern safety devices of today.

Thus it was on Saturday 12 April, when Wardell was making his customary visit to his office, that he had a serious accident. According to a report in the *Advocate,* reprinting a similar one in the *Freeman's Journal* of April 16, he was in the act of leaving the lift 'which usually remains stationary' when the contraption started its descent without any warning. '... the gentleman was struck on the back and thrown backwards, one leg being jammed against the landing.'[1] A doctor was sent for and on arrival found that Wardell's right thigh had been fractured in two places. 'The injured gentleman was subsequently removed to his home at North Shore'.[2]

We can imagine a difficult, painful return to Upton Grange by way of carriage and ferry, and a lengthy period of enforced inactivity until the injured leg was healed. The Melbourne *Advocate* report concluded by reminding its readers that Mr Wardell was the architect of two great cathedrals and it hoped that he would make a full recovery in time to be present at the celebrations in connection with the consecration of the completed St Patrick's Cathedral, and that 'it will fall to our pleasant duty to record the fact of Mr Wardell's recovery from the painful accident which has elicited so much sympathy from his friends in our own colony of Victoria.'[3] As the consecration of St Patrick's was not to take place for another five years, the newspaper's hopes for his recovery were most likely to be realised. But whether it would be a 'full recovery' remains open to speculation. A serious accident of that nature occurring at Wardell's age was bound to have a degenerative effect on his health. Judging from his correspondence, particularly the noticeable deterioration of his handwriting, there appeared to be a marked change in his general disposition.

He had recovered sufficiently by the middle of May to supervise tenders called for additions to the Citizens Life Assurance Company Building, and from August to September was corresponding regularly again with the Archbishop in Melbourne on the subject of furnishings for St Patrick's in which he took a proprietary interest. In November and December he turned his attention to a proposal to remove the organ and choir in St Mary's from the chamber designed for them, and place them at the south end of the nave. 'I venture to express my hope that it may never be permitted.'[4] Today when most visitors enter the Cathedral through the popular, preferred entrance, the west door fronting College Street, they pass under the organ gallery and through an inner door and marble barrier. Wardell argued against this as far back as December 1892 when he wrote that if the organ gallery had to be moved, the East Transept should be chosen and he gave his reasons,

'1.   It would not then dominate the altar of the Blessed Sacrament or interfere with the worshippers.

2.   It would not cause that very unpleasant sensation, which I think all of us experience in entering a large building from under a low ceiling which is best described by the ordinary expression, "feeling crushed".

3.   It would not destroy the view of the Tower Arches on entering the building by the principal doorway.'[5]

Among the collection of Wardell's business letters which survive from those last years, two especially may be noted as revealing different aspects of his character, and which seem at first reading to be mutually contradictory. The first of them advises the Archbishop in Melbourne that he is reducing the wages of Mr Campbell, the Clerk of Works, from £6 per week to £4. He explains to Carr that he does not think he is justified in continuing to pay the higher sum 'In view of a general reduction of wages.'[6] To us this seems a mean and insensitive action possibly resulting in hardship for Mr Campbell. As Clerk of Works at St Patrick's he had the onerous responsibility of the management and supervision of operations at the building site. Our contemporary understanding of social justice and the need for a just wage is offended by Wardell's action. Suddenly the image of Wardell as a Scrooge-like employer, comes to mind.

Our judgement, however, should be tempered by looking at the economic conditions at the time. Victoria was the State most heavily affected by the notorious Bank Crisis of 1893. At that time the banks, building societies and financial institutions failed owing to many factors following the great strike of 1890 which had led to distrust in the minds of depositors and investors. Large holdings were withdrawn and hoarded. Much capital was locked up in the failed banks and not distributed as wages, resulting in widespread unemployment. Wages fell precipitately as did rents. 'The change of fortune proved disastrous not only to many families previously to all appearances in opulent circumstances, but by all classes alike their reverses were borne with the greatest bravery.'[7] In spite of a widespread cessation of building work generally, St Patrick's Cathedral and St Mary's in Sydney managed to survive. In these circumstances Wardell, like other employers, was merely responding to economic pressure, but in his case managing to maintain employment where others around him failed. He regarded his action as a 'simple matter of duty' - his duty to the Archbishop and the Cathedral to cut costs and avoid a greater hardship.

How, we may ask, did the builders and craftsmen employed on Wardell's projects regard him, in view of this and presumably other associated economies? Did Campbell harbour any resentment? Evidently not; he was one of those who bore his reduced circumstances 'bravely'. Wardell's workers held him in deep affection and with great respect if the second letter to which we refer is any guide.

Mr Campbell resigned as Wardell's Clerk of Works some eighteen months after his wages had been reduced, the elapsed time seeming to suggest that the reduction was not the cause of him leaving. In his letter of resignation he writes: 'I cannot omit to thank you heartily and sincerely for many an act of kindness that you, dear Sir, have bestowed upon me in the eleven years I have had the honour to be in your service.'[8] Wardell acknowledges the resignation 'with regret' and adds, 'I am sorry to part company [with you] and I sincerely wish you success which I feel sure you deserve.'[9] Wardell was not an employer who habitually cut others' wages while maintaining a high return for himself. In his dealings with the Building Committees of both Cathedrals there were instances when he made concessions in the matter of fees. In one letter he agreed to waive his 2½% and 'accept as an honorarium, whatever you may propose.'[10] Prominent architect historian, the late Charles Glanville, in his authoritative essay on Wardell and the building of St Mary's argues that the architect was, in fact, paid a pittance for his work on the Cathedral.[11]

Towards the end of 1893 Wardell was advising on stone carvings for St Ignatius's Church in Richmond, and in January the following year made another of his many visits to Melbourne. By this time he had evidently recovered sufficiently from the accident nine months previously. His main concerns on that visit were the revised Health and Safety Regulations insisted upon by the Victorian Board of Health in respect of St Patrick's. He complained that the Church was commenced and partially occupied before the regulations were issued. The new regulations demanded that 'the total egress space should be equal to half the square root of the number of persons, allowing four square feet for every person accommodated.'[12] The number of persons to be considered was, by the Board's ruling six thousand, nine hundred and fifty. Wardell submits the plans to the Board which show that the numbers accommodated were more like two thousand, four hundred and fifty-two persons, allowing 1½ feet seating space for each person in the benches, and the benches spaced 2½ feet apart from back to back. This had been accepted by the Board originally but they now demanded changes to the plans. The argument continued over the next two years. The Board, in making its complaints, did not limit its objections to the amount of egress space, but insisted upon greater ventilation at clerestory level. We prefer to think of Wardell the engineer grappling with structural problems, or the artist bent over his drawing board creating inspired design features, but as both Cathedrals were nearing completion and partially in use, much of his dwindling energy was spent in fire prevention installations, safety features, advising the archbishops in countless letters about lightning conductors, arguing with the Sydney City Council over plans for realigning the perimeter roads and the effect this would have on the length of the Cathedral nave. He was so frustrated at one point that he asked Archbishop Carr to urge a friendly member of parliament in Victoria to keep his eyes and ears open when the new Health Act was debated. 'Watch the new Bill and perhaps check the preposterous powers given to the Board by the old Act.'[13]

The view of St Mary's in the mid 1890s, when approached from around St James' Road at the north end of Hyde Park and the south end of Macquarie Street, would look much as it does today. The massive north end, with its flying buttresses, the clerestory windows above the chancel and the roof, were all finished. The trees of Hyde Park would have screened the unfinished central tower and the forest of wooden scaffolding enfolding the incomplete and uneven walls of the nave. Instead of the fast cars and commercial trucks of today, buses speeding around the corner and parties of tourists and office workers waiting at pedestrian cross-ways, the traffic in the last decade of the nineteenth century was light and consisted of the occasional tram, horse-drawn carts, and a line of cabs – horse-drawn taxis - awaiting custom beside the road. This was the quiet scene that Wardell would have known when he visited the site on most days. He took a paternal interest in progress, consulting with the Contractor and the Clerk of Works, and whenever the Cardinal's engagements permitted, calling on him personally. He had insisted on dealing directly with His Eminence on matters of expenditure and the choice of furnishings. Often his letters confirmed their earlier discussions and the decisions they had agreed upon. If they were unable to meet, Wardell was meticulous in keeping Moran informed by letter. His handwriting at this time had deteriorated seriously and is often difficult to decipher - small, and scrawly - but his grasp of detail and his concern that the highest standards be maintained were as important to him as always.

Having submitted his designs for the High Altar to the stone carvers, Farmer and Brindley of Westminster, he writes later to say that he is not satisfied with the detail of the alabaster bases of the small columns, 'I enclose an amended section which I will be glad if you will substitute for the first [designs] and make the bases octagonal instead of circular.'[14] The splendid, massive altar with its rich embellishments and carvings is no longer used in today's liturgies and is too distant to attract the interest of visitors. It is, however, a striking work of Gothic design consisting of a stone table supported by Wardell's five marble columns with capitals in carved alabaster, and behind them, panels of richly carved bas relief sculptures depicting the Last Supper and the Crucifixion. Above it is the majestic white marble reredos that remains a central point - an attraction for the eye - in the Cathedral today.

Wardell had witnessed experiments to test a liquid composition applied to wood to deter attack by termites and assist fire prevention. He advised the Building Committee to invest in it. 'I strongly recommend it if only for the satisfaction of feeling, in the event of any unfortunate accident happening hereafter that they had done their best to prevent it and were not liable to any reproach.'[15]

In the last year of Wardell's long life, the letters to both the Archbishop of Melbourne and the Cardinal in Sydney continued unabated. He was anxious for the completion of the central Tower of St Mary's, popularly known as 'the Cardinal's Tower.' 'I have still every hope that the close of the year will see its completion. I have however requested the contractor to go on also with the north wall of the West Transept to the same level as the south wall. Fixing the roof of the East Transept is now in progress.'[16]

In March 1899, the year of his death, he reported to the Cardinal that the hanging of all the bells in the Tower 'is now satisfactorily completed'.

He wrote to Dr O'Haran, the Cardinal's Secretary, explaining why it was important that the central Tower should be completed to the height that he had determined.

> 'It is placed at the very heart of the work, and its walls grow out of it and form part of it. The building therefore of any portion of it will have this effect of raising or heightening the whole structure. This may be judged even now by looking at the building from its western side, and realising what contrary effect would happen if the small portion of the tower now built was not there.'[17]

In mid 1899 he is concerned with the designs and with commissioning Farmer and Brindley for altars for two chapels, the Blessed Sacrament and the Sacred Heart. It is clear that at the time of writing he had some intimation that his life was drawing to a close. Although his son was later to write that his father's death was sudden and unexpected, who can say what inner weakness Wardell was secretly conscious of, but which he kept only to himself? In support of this view there exists a premonitory, somewhat ominous letter to Dean O'Haran on the subject of the altar plans, 'I hope I may live to give the order myself, that I may accompany it with all such further particulars and details, but if I do not, I believe you will be practically safe in sending the drawings you have to them.'[18]

A letter to the Cardinal dated September 15, contains instructions for the inspection and cleaning of objects in the Cathedral. 'The brass railings around the Sanctuary, the Standard Lights, and the Communion Rail, require cleansing and re-burnishing. This is a work that should not be delayed as delay adds to the difficulty and the cost of it.'[19] He recommends that an expert be appointed to inspect and maintain the gas lighting.

His final letter on Cathedral business is to Dr O'Haran, the Cardinal's close friend and secretary, and is written from his office in George Street six days before his death. 'I return to you the plans of the southern half of the Cathedral which I borrowed for reference in making a plan for the approaches to the southern front and I have added this plan to the others'. Wardell then adds a warning showing how anxious he is that the plans are not lost, 'they are valuable documents and are likely to get lost or scattered without special care.'[20]

There are no extant letters from him after November 13, and one assumes that he may have been confined to the house. There is little doubt that he was ailing at this time in spite of his son Herbert's contention that his final illness was of less than twenty-four hours duration. His condition, later diagnosed as pleurisy, worsened over the next few days.

William Wilkinson Wardell received the last rites of the Church from Father Joseph Brennan and died peacefully on Sunday evening November 19, aged 76 years. 'He was conscious to the last' and was surrounded by three of his surviving seven children. Edward was alerted to his father's condition but had not yet arrived from Melbourne. Herbert Edward had been summoned from his home in Ocean Street, Woollahra. The youngest and unmarried daughter, Constance, his companion and housekeeper, was there; so too were his married daughters, Kathleen Mary Power and Ethel Mary Covely. Family friend, Dr J. H. Kyngdon, attended his patient and signed the Death Certificate with the simple explanation that he died of pleurisy leading to heart failure. The Certificate contains one known error, perpetuating the fiction that William's father, Thomas Wardell, was a man of 'Independent Means'. Thus, it was hoped, the great architect and engineer's early life would, like Shakespeare's bones, remain undisturbed.

Wardell was buried two days later next to his wife, Lucy Anne and his son, Lawrence George, in the family grave in the Catholic section of what is now known as Gore Hill Historic Cemetery on the North Shore. Coincidentally, in a less conspicuous grave a few paces away, would lay, nine yeas later, the body of Australia's only beatified saint, Mary McKillop. Wardell, who lived almost next door to her convent, surely would have known her. [21]

When the Cardinal was informed of Wardell's death he suggested a public tribute be paid to him, with an Episcopal Mass and burial from St Mary's Cathedral. But the family, faithful to Wardell's own wishes, chose instead a private and little-publicised local ceremony. The Cardinal was represented by Dr O'Haran and several priests from the Cathedral and local parish. The coffin was carried to the graveside by four of Wardell's closest associates: the then Government architect and former business partner, William Vernon; his close friend, architect J. Horbury-Hunt; John Reid, Clerk of Works at the Cathedral, and Thomas Loveridge, the Contractor. His son, Herbert, wrote to the Dean of St Patrick's Cathedral with the news and followed with a second letter to thank him for his immediate expression of sympathy: 'My father's death must have been a very great shock to you and also to His Grace the Archbishop.'[22]

Tributes and fulsome obituaries followed in the ensuing weeks. The *Sydney Morning Herald* led the way by announcing to the world at large the death of 'one of the oldest and most eminent architects of Sydney'. The paper claimed that he had been ailing some time. The obituary writer regretted that Wardell did not live to see the completion of his 'splendid design' but in a flight of poetic fancy suggests that, 'For generation upon generation, the memory of him will be perpetuated by praying hands, by the eternally pointing supplicating hands and spires.'[23]

The *Freeman's Journal*, the Irish Catholic paper, was the most expansive in its praise of him:

'He closed his life of noble labours, a life crowded with artistic triumphs in a manner in keeping with the modest gentleness which marked his whole career. He had devoted the best of his rare gifts to the service of religion. His name will go down in history as the man who designed and directed the erection of two great Cathedrals making him a truly remarkable figure among architects not of Australia alone, but of the world.'

The paper concluded with a telling paragraph, a confirmation of the character we have gradually got to know:

'No man was ever held in higher respect by his friends, and yet his intimate friends invariably addressed him by a pet name [Will, or Willy]. It may need, perhaps, the writer's memory and associations to find this a touching expression of his winning character, his playful smile, and pleasant ways.'[24]

The London Catholic journal, the *Tablet*, noted his death with a long summery of his life in their edition of January 6, 1900. Their report was closely based on information passed to them by 'one of his sons' - although whether the source was the eldest (Edward), or the youngest (Herbert), is not revealed. Much of the phrasing appears in all subsequent biographical notes so that little is revealed of his early life, merely the recurring phrase, 'he was educated for the profession of civil engineer,' and that he had a 'strong desire to be a sailor at an early age and went to sea.'[25]

On the principle of De mortuis nihi nisi bonum, the *Age* in Melbourne, his old adversary, noted the death of the 'prominent architect' two days later. Wardell's contribution over many years to the grandeur of Melbourne's architecture and his tireless civic work is dismissed with the brief sentence: 'Between 1853 and 1878 he was connected with the Victorian Public Works Department.' The *Age* waited over thirty years to make full restitution for its earlier attacks on him. A much belated obituary-type article appeared over the initials H.W.L.S. in 1936, headed 'A Noble Architect'. It spoke of his fertile and prodigious capacity for work and how, under his administration, the Department [Public Works] reached a high standard of efficiency.

'The city was fortunate indeed that both as a creative artist in stone and timber and as a fecund and original planner of all types of public buildings he was probably the equal of any architect of his day. To all antagonists, whatever the provocation, he was always the polished gentleman.'[26]

From mid-year 1899 there is considerable evidence that Wardell realised his death could not be far distant. Not only is there a hint of urgency in his prodigious flow of letters concerned with the designing and ordering of furnishings for his cathedrals, as if he were attempting to leave nothing to chance - we have already noted his ominous phrase, 'I hope to live long enough to give the order myself,' - but in a more obvious sign of preparation, he drew up his will and dated it June 26 of that year in the presence of his Executors. From this we learn that, after sundry debts were paid, his estate was valued at £12,919 – considered a reasonable fortune in the values of those times. The document itself, handwritten, runs to five pages, is detailed, excessively pleonastic in the manner of legal documents of the time, but precise in its intentions. He leaves all his office furniture fittings, specifications, contracts and plans, to his younger son, Herbert Edward, and hopes that 'in the case of him being employed in any development or continuation of my works he will consult and ask the assistance of my faithful clerk and assistant, George Denny and remunerate him fairly.' Wardell excludes from the furniture and fittings his 'writing table and the larger of the two Milner safes which are the property of my eldest son, Edward Stanfield Wardell.'

To his unmarried daughter, Constance, Wardell bequeaths all his linen, his china and glass, also his pictures and prints, his works of art and ornaments. There is provision, also, for his two married daughters to choose items from his collection of jewellery, 'and such articles of my personal effects as they may choose for their own use only 'and not for giving to others except by way of legacy.'

When Wardell comes to the dispersal of his money he leaves small sums in cash, £500 to Constance, and to his eldest son, Edward, £1,250 and to Herbert, the sum of £250. There are various other sums to his married daughters, and £100 to George Denny, his clerk and a small sum to the local church. The major portion of the estate was to be re-invested and managed by the Executors, Thomas Dibbs and Thomas Gaden of Sydney, and the income apportioned to members of his family - the greater part to Constance, the unmarried daughter, and the residue split between Edward, Herbert, Ethel and Mary. Constance was to inherit the lease of Upton Grange.

Apart from the natural and sincerely held sorrow at losing a faithful advisor, Wardell's death presented the Building Committees of both unfinished Cathedral's with practical difficulties. Both St Patrick's and St Mary's were fully his creations. Like a loving parent he watched over them, monitored their progress, and guided and advised all those who worked on them including Archbishop Carr and Cardinal Moran. He protected his Cathedrals from ill-advised or harmful alterations to his designs, and his refined taste and sensitivity rejected shoddy work or cheap artefacts. The feeling at the time – so often repeated in similar circumstances – was surely that Wardell was indispensable and nobody could adequately take his place.

But he himself had made provision for the inevitable.

In his will Wardell ensured that all his papers, correspondence, plans and notes remained in the hands of his firm, Wardell and Denny and that his son, Herbert, now a partner in the firm, would be enabled to take his place and supervise the completion of his works. Henceforth the Wardell of Wardell and Denny would be not William Wilkinson, but Herbert Edward.

The traditional service he had established would continue, but architectural styles were about to change with the dawn of the twentieth century. The Gothic Revival had run its course. Architects working in concrete and iron were spreading their gospel from America; functionalism and Art Nouveau were born.

Never again would a cathedral be built in the Gothic style; historicism was dead. The modernist revolution had begun and would gather momentum in the first half of the twentieth century, supported by the liturgical changes of the 1960s. Concrete brutalism was its natural heir. Cathedrals designed in this style, typified by Christ the King in Liverpool, England; Saints Peter and Paul, Bristol, and Our Lady of the Angels in Los Angeles and many similar-style smaller churches, would become the dominant fashion - described by architect and critic, Moyra Doorly, as 'the ugliest and emptiest churches in history.' [27]

Perhaps we may say in conclusion that William Wilkinson Wardell died at the right time. He had been born at the dawn of the Gothic Revival, and died when Gothicism had run its course. He now rests in peace, knowing nothing of the architecture revolution that would overtake - but happily not diminish - his life's work. In the eyes of the multitudes who have come, and will still come, to St Patrick's and St Mary's Cathedrals to admire and to pray, their stones remain a beacon of Faith, a living gospel.

Wardell's two great cathedrals, together with his many other churches can never be described as Moyra Doorly describes post-modern churches designed and built following Wardell's death, as 'the ugliest and emptiest in the world.'

Upton Grange, the Wardells' family home on the North Shore. Photograph taken in 1881 showing William Wardell on the veranda admiring his son, Herbert, as he prepares to ride on his horse to St Ignatius' College, Riverview. Copyright, Mitchell Library.

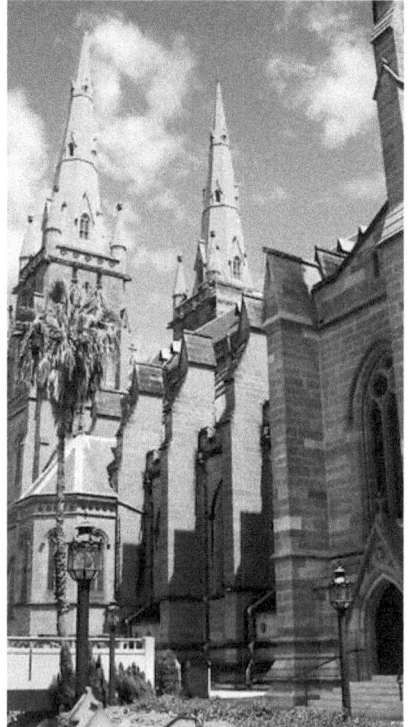

Views of St. Mary's Cathedral, Sydney.
TOP: From the north, the 'Cardinal's Tour' dominating. LEFT BELOW: Another view of the massive tower. RIGHT: The two spires added in the year 2000.

Lincoln Cathedral (right), and St Mary's Cathedral (below). Views which show how Wardell may have had Lincoln in mind when he designed the north end of St Mary's. They also show how Wardell avoided over-decoration and too close a comparison, preferring to adapt his design to Australian requirements while preserving essential elements.

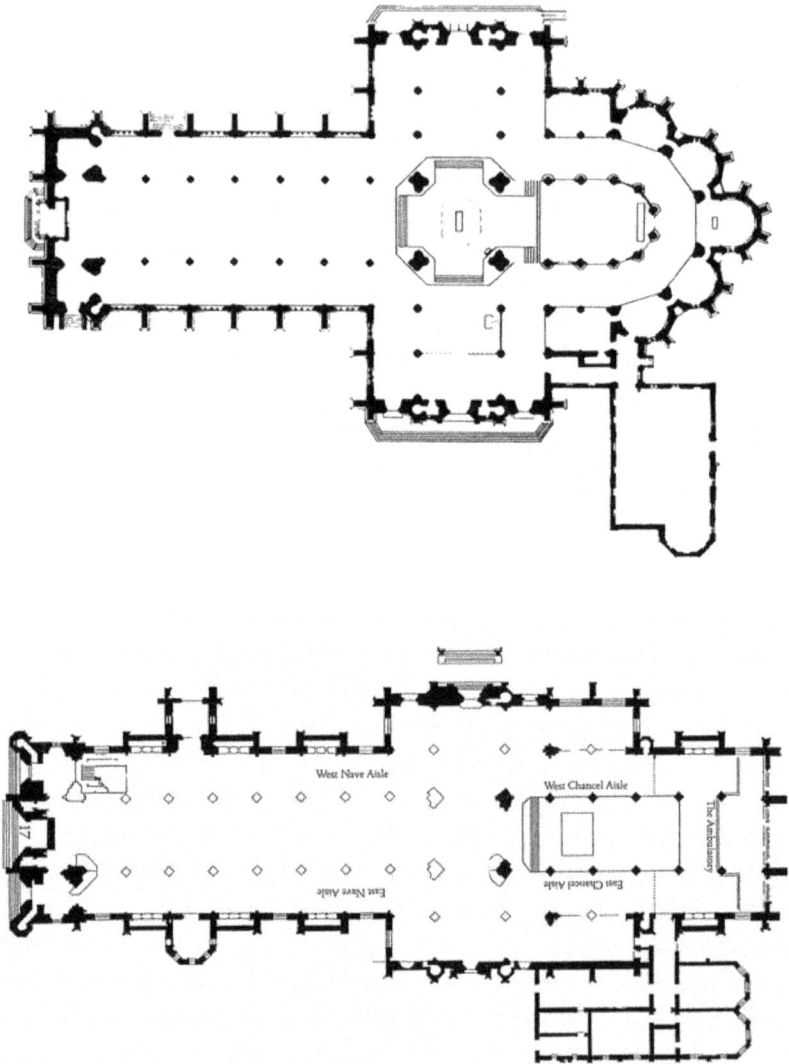

A comparison of the ground plans of St Patrick's Cathedral (top) and St Mary's Cathedral (below) The nave of St Patrick's is wider having seven bays before the crossing. St Mary's is narrower but longer having eight bays.

258

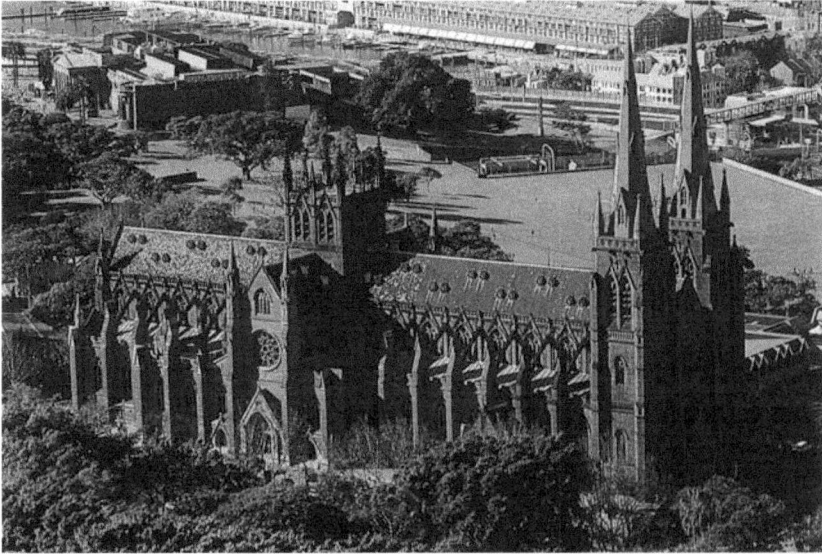

(Top): St Mary's Cathedral, general view from the south-west with the Domain beyond, and the trees in Hyde Park in foreground.   *(Copyright, St Mary's Cathedral)*
(Below): The central tower over the Crossing once popularly known as 'the Cardinal's Tower', so named after Cardinal Moran.

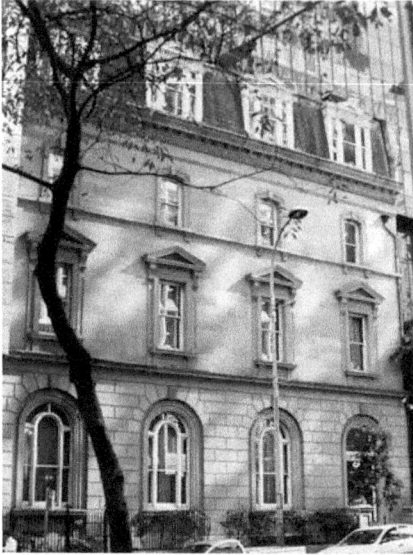

(Left) Wardell's former New South Wales Club as it appears today, a brave survivor surrounded by steel and glass high rise offices in Bligh Street in the heart of the city.

(Below) The handsome Grafton Bond Warehouse on Hickson Road. The bands of deeper-toned brickwork, the arched capping of the windows, and the roof gables with their little windows relieve what otherwise might be a too-stark elevation.

Two views of St John's College within the University of Sydney designed and building commenced shortly after Wardell arrived in Australia in 1858. Delays and poor communication between architect and client, and accusations on both sides, led to Wardell's resignation before the building was completed.

261

TOP: Wardell's old home in North Sydeny, Upton Grange, as it appears today part of Shore Grammar Junior school in Edward Street. The house was leased by Wardell, but not designed by him.
BELOW: St Michael's House, the first permanent school building at Riverview Jesuit College. Designed in 1880 by Wardell who was Consultant Architect to the College for a short time. Wardell was dismissed after he had complained that the bricks were of inferior quality. The addition of cement rendering at a later date shows how right he was.

Two views of the former offices of the Australian Steam and Navigation Company on Hickson Road in the Rocks area. Here is Wardell at his most imaginative, adapting Dutch influences evident in its stepped gables, and a steeple and castle-like tower, all of which add to its visual delight. Much loved and well-preserved, it has become a tourist attraction housing an art studio and craft shops. Once a prominent feature on the western arm of Circular Quay much of it is now hidden behind the Passenger Terminal.

Wardell – 'a position of eminence'.
Copyright: MDHC Catholic Archdiocese of Melbourne.

Wardell, the eminent architect in old age, in his library.
Courtesy MDHC Collection.

APPENDIX 1

Extant buildings wholly or in
part designed by William Wardell

Dating churches is often confusing on account of several dates in the records – those relating either to the commission, the plans, laying of foundation stones, building commenced, or blessing and opening, or later additions. Dates given below are when work commenced, unless otherwise stated.

ENGLAND & SCOTLAND:

1. **Abingdon, Berkshire.** Church of St Mary & St Edmunds. Original scheme for church and presbytery only partly realised. Formally opened 30 September 1857. Chancel, south aisle and sacristy Wardell's. Completed by George Goldie.
2. **Brentwood (Nr.)** Thornton Hall. Chapel erected for Lord Petre and completed 1857. Now in hands of Brentwood Diocese.
3. **Brook Green, Hammersmith.** Church of Most Holy Trinity, Presbytery & Almshouses. Opened 26 July 1851. Tower and spire by J. A. Hansom added 1871.
4. **Chislehurst, Kent.** Church of St Mary and also Presbytery. Opened 1854. Chapel added 1871.
5. **Clapham, SW4.** Our Lady of Victories. Foundation stone Aug. 1849. Opened 14 May 1851. Enlarged with Lady Chapel and Transept by J. F. Bentley.
6. **Dorchester-on-Thames, Oxford.** St Birinus. Opened 1849. (Cf Tablet, 28 Aug. 1849)
7. **Greenwich, Grooms Hill, SE10.** Our Ladye Star of the Sea. 1849. Suffered damage during the war. Restored but 1965 interior modifications now much regretted.
8. **Galashiels, Selkirk.** Our Lady & St Andrew. Opened 2 Feb. 1858. Full length of nave not realised owing to lack of money. Later additions and reopened 1878.

9. **Hampstead.** St Mary's, Holly Place. Only the Italianate West Front is by Wardell, designed and erected in 1850.
10. **Kelso, Roxburghshire.** St Mary of the Immaculate Conception. Opened in 1858 after departure of Wardell.
11. **London, Commercial Road, E.1** SS Mary & Michael. Foundation Stone 1853. Opened 1856. Severely damaged in WW11 & since repaired with alterations.
12. **Liverpool, Mount Vernon St.** Conventual Church of Our Lady of Mercy. Opened 20 September 1857.
13. **Manchester, Bedford Street.** Church & School of St Wilfred. Opened 1850.
14. **Richmond Mechanics Institute.** 1843. Probably the first building designed by Wardell when he established his practice, Wardell and Littlewood, in the city of London. Italianate in style and can still be seen in Richmond,Surrey, but the appearance and interior are much changed today.
15. **Upper Norwood, Central Hill S.E.19.** The 'Wardell Wing' of the Convent only. 1855-57.
16. **Wisbech, Cambs.** Our Lady & St Charles, Foundation Stone Nov. 1853. Opened 1 Sept. 1854. Clumsy, inappropriate tower and other changes added in 1962.

VICTORIA: CHURCHES

1. **East St.Kilda.** St Mary's and Presbytery, Dandenong Road. Original, smaller church 1858, Redesigned and enlarged 1870.
2. **Warnambool:** St Joseph's 1863.
3. **Williamstown:** St Mary's 1859.
4. **Wangaratta,** St Patrick's, Ford Street. 1863-1865.
5. **Geelong.** St Peter & Paul's, Ashby, now West Geelong,1861-1864.
6. **Hamilton:** St Mary's,1864-1865.
7. **Heidelberg:** St John's, Yarra Street,1859.
8. **Koroit:** Infant Jesus, 1866.
9. **Kyneton:** Our Lady of the Rosary, 1862.
10. **Melbourne.** St Patrick's Cathedral, 1858-1897.

11. **Richmond.** St Ignatius, Church Street, 1867-1892.
12. **Toorak.** St John's Anglican Church. 1860-1865.
    Tower and spire built in 1870.

VICTORIA - BANKS AND OTHER BUILDINGS

1. **'The Gothic Bank'** Cnr. Queen and Collins Streets now the ANZ. 1883. Unique Italianate Gothic design on which no expense was spared. Cost estimated at just over £77,000. Other banks which survive include one at Camberwell Victoria, and two as private houses - one at Moonee Ponds and another in North Melbourne.
2. **Government House:** 1871-76. Among all the public buildings in Melbourne supervised by Wardell, this is the only one which we know to have been wholly designed by Wardell himself.
3. **Genazzano College for Girls:** Cotham Road, Kew. Commissioned in 1889 by the Sisters of the Faithful Companions of Jesus (FJC). Original building completed in 1891.

NEW SOUTH WALES: CHURCHES

1. **St Mary's Cathedral:** Building of the third church on the site to Wardell's design began in 1868. Building continued throughout Wardell's remaining lifetime and beyond. Although partially occupied during his lifetime and a dedication ceremony was held on September 8, 1882, the Cathedral was not finally completed until 1928, 29 years after the architect's death. The spires were added in the year 2000.
2. **Lismore.** Claims that St Carthage's Cathedral was designed by William Wardell are mistaken. In 1892 he may have contributed advice and early sketches for a design, but work was shelved. Not until 1904, five years after his death, were the foundations laid and building work begun. The commission was given by Bishop Doyle to Wardell and Denning and the work was carried out by Herbert Wardell who inherited his father's practice. The building was completed and consecrated in 1919.

NEW SOUTH WALES: BANKS AND OTHER BUILDINGS

1 **St John's College,** University of Sydney. Missenden Road. Although designed by Wardell, and building commenced in 1858, Wardell resigned as architect in 1859. Edmund Blackett was engaged to complete the work faithful to Wardell's design.

2. **Australasian Steam & Navigation Co.** Offices and warehouse in Grafton Street, the Rocks, Sydney.

3. **NSW Club.** Bligh Street, Sydney.

4. **Grafton Bond Warehouse,** Hickson Road, Sydney

5. **Banks:** Wardell designed around one dozen banks for the English, Scottish and Australasian Company, and the New South Wales Banking Corporation. Many of these have been replaced by modern buildings. A happy survival is the well -preserved bank dating from 1884 found in the Southern Highlands country town of Berry. Here the stepped gables, oriel window in the gable, and the stone capping over the lancets are typical Wardell features of many of his banks. The building is now a museum run by the Berry and District Historical Society.

TASMANIA

Hobart. St Mary's Cathedral which was demolished because of faulty foundations - the responsibility of a local builder. Later rebuilt incorporating some of Wardell's design.

APPENDIX 2

Wardell's memorandum listing the volume of work in the PND Melbourne

APPENDIX L.

MEMORANDUM FOR THE HONORABLE THE MINISTER OF PUBLIC WORKS.

In accordance with Mr. Fraser's desire, I have the honor to submit herewith a list of the several works and contracts in progress at the commencement of the financial year '72-'73 and those entered into during its currency by this department, and which are being executed under its direction.

In addition to such works as those on the River Yarra, dredging dock entrance, clearing the River Murray, &c., it will be seen that, during the financial year referred to, we have entered into 138 new contracts, varying in amounts from £100 to £64,500, and into 210 new contracts under £100 in amount.

Several contracts of former years have also been in progress.

As the state of the works of the graving dock are perhaps now of the most interest and importance, I would report that the masonry of the dock proper is practically completed, as all that remain to be done are the draining culvert, timber and boat slides, and south wing wall. These will be finished in six weeks. Tenders will be invited during the next week for the boiler and engine house, the dock offices, gatehouse, &c.

The caisson will not, I am of opinion, be ready before the end of September; but I know of nothing likely to prevent the dock being ready for work early in November, as the pumping machinery is in a very forward state.

Of the other works I need only remark that, with one exception, they are progressing satisfactorily : the exception is the harbor works at Belfast, where the contractors have stopped a portion of the work on the ground of dissatisfaction with the measurements. We are acting in the matter under the advice of the Crown Solicitor, and the contractor has been warned that the conditions will be strictly enforced.

18th June 1873.   W. W. WARDELL.

			£	s.	d.	
72.3/1.		Dredging Operations, and other River and Harbor Improvements in the River Yarra and Port Phillip, including the Maintenance and Repairs of Steam Dredges and Vessels in connection therewith.	13,000	0	0	
2.		Plant, and other expenses for landing Silt	7,000	0	0	
3.		Clearing the River Murray	5,300	0	0	
68/57.	Williamstown	Alfred Graving Dock	118,533	12	6	
71.2/103.	Kew Asylum	Boundary Walls, Outbuildings, &c.	23,268	8	3	
71.2/82.	Williamstown	Pumping Machinery, &c., Graving Dock	8,150	0	0	
71.2/214.	Kew Asylum	Gas and Water, &c.	7,575	13	4	
71.2/3.	Williamstown	Caisson for Alfred Graving Dock	11,326	0	0	
71.2/195.	Melbourne	Stone Wharf Extension, south side	4,453	0	0	
72.3/2.	Williamstown	Repairs to Hull, &c., Dredge *Griper*	2,653	10	0	
3.	Melbourne	Alterations to Engines, Mint	268	0	0	Completed.
4.	Williamstown	Fifty Pontoons for Dredging	7,300	0	0	
5.	Toorak	Works for Birthday Ball	370	0	0	Completed.
7.	Melbourne	Works at Immigration Depôt	110	0	0	Completed.
9.	Sandhurst	Cementing, &c., Post and Telegraph Office	310	0	0	Completed.
11.	Melbourne	Additions to Land and Survey Office	457	15	2	Completed.
13.	Melbourne	Piling, &c., Gas Dock	129	10	0	Completed.
14.	Sunbury	Removal and Erecting Iron Store	406	5	3	Completed.
15.	Melbourne	Additions, Housekeeper's Quarters, Parliament Houses	149	0	0	Completed.
19.	Williamstown	Caulking H.M.C.S.S. *Nelson*	195	19	6	Completed.
20.		Twenty Portable Buildings for Police	1,646	0	0	Completed.
21.	Majorca	Fencing, &c., Court House	112	0	0	Completed.
22.	Melbourne	Iron Gates, General Post Office	479	0	0	
23.	Williamstown	Metalling Approach to Ann-street Jetty	161	0	0	Completed.
24.	St. Leonard's Jetty	Removal of Hulk and Piling	253	18	0	Completed.
28.	Castlemaine	Repairs at Court of Petty Sessions	139	0	0	Completed.
38.	Smythesdale	Fencing Public Buildings Reserve	189	10	0	Completed.
41.		Round Telegraph Poles, Telegraph Department	197	7	11	
44.	Melbourne	Works at Industrial Schools	165	5	0	Completed.
54.	Murray River	Snagging Punts	1,252	0	0	
56.	Wood's Point	Works and Repairs to Court House	249	0	0	Completed.
57.	Melbourne	Works at Wharves	35,796	0	0	
58.	Geelong	Works at Industrial Schools	1,750	0	0	Completed.
64.	Melbourne	Fencing, &c., Treasury and Parliament Reserve	119	0	0	Completed.
67.	Emerald Hill	Works at Post and Telegraph Office	149	7	6	Completed.
74.	Majorca	Removal, &c., Court House and Police Buildings	329	10	0	Completed.
75.	Kilmore	Alterations and Repairs to Gaol	265	2	0	Completed.
76.	Melbourne	Cast-iron Pipes for Sewerage and Water Supply Office	14	10	0	per ton.
77.	Talbot	Stable, Police Station	151	0	0	Completed.
81.	Geelong	Repairs to Gauging Shed, Customs Reserve	184	10	0	Completed.
84.	Avoca	Post and Telegraph Office	1,088	10	0	
85.	Kyneton	Additions to Court House for Sub-Treasury	324	19	6	Completed.
86.	Melbourne	Works at Telegraph Office, General Post Office	135	0	0	Completed.

			£	s.	d.	
87.	Melbourne ...	Works at Land and Survey Office ...	197	9	0	Completed.
88.	Ballarat ...	Works at Gaol	965	0	0	Completed.
89.	Mordialloo Creek	Snagging	168	0	0	Completed.
92.	Portarlington Jetty	Shed	607	6	0	
93.	Kew Asylum ...	Furniture	1,423	10	6	Completed.
94.	Sandhurst ...	Alterations to Post and Telegraph Office ...	847	0	0	Completed.
97.	Melbourne ...	Works at Industrial Schools, La Trobe street	546	19	0	Completed.
103.	North Carlton...	Fencing Water Reserve Holo ...	218	12	0	Completed.
105.	Melbourne ...	Government House	64,547	12	0	
106.	Melbourne ...	Works, Housekeeper's Quarters, Houses of Parliament	124	10	0	Completed.
107.	Wood's Point ...	Repairs, Warden's Office	127	0	0	Completed.
108.	Kew Asylum ...	Repairs to Detached Lodges	129	7	0	Completed.
114.	Williamstown ...	Repairs to Dredge *Wombat*	213	0	0	
119.	Koroit ...	Works at Court House	175	18	0	Completed.
120.	Warrnambool ...	Works at Gaol	140	2	9	Completed.
126.	Williamstown ...	Repairs to Dredge *Alligator*	963	15	4	
130.	Melbourne ...	Fittings and Furniture, Treasury ...	123	10	0	Completed.
131.	Williamstown ...	Filling in Stone, &c., Dock Yard ...	1,105	19	3	
		And Rates Gazetted	7,975	19	3	
136.	Yan Yean ...	Works to Tunnel	747	15	6	
135.	Melbourne ...	Fire Extinguishing Apparatus, General Post Office	135	10	0	Completed.
138.	Clunes ...	Additions to Post and Telegraph Office ...	115	4	0	Completed.
139.	Beechworth ...	Water Mains to Asylum	1,435	0	0	Completed.
141.	Williamstown ...	Stone Filling, Steamboat Pier ...	410	6	8	
142.	Ballarat ...	Fencing Gaol Reserve	162	7	0	Completed.
145.	Stawell ...	Stables for Police	298	0	0	Completed.
147.	Melbourne ...	Works at Victoria Barracks	155	4	0	Completed.
153.	Maryborough ...	Repairs to Police Quarters	152	0	0	Completed.
154.	Kyneton ...	Repairs to Police Station	114	13	6	Completed.
158.	Wangaratta ...	Police Stables	305	0	0	Completed.
160.	Kew Asylum ...	Hydraulic Lift	278	0	0	
162.	Macedon ...	Fencing State Nursery	112	6	6	Completed.
167.	Yarra Bend ...	Works at Asylum	185	0	0	Completed.
168.	Williamstown ...	Completion of Stone Embankment, River Yarra	1,439	0	0	
170.	Sunbury ...	Stone Wall Fencing, Industrial Schools ...	222	15	0	
171.	Belfast ...	Harbor Works	1,688	0	0	
173.	Sunbury ...	Works at Industrial Schools	433	0	0	Completed.
174.	Melbourne ...	Iron Tubular Sub-Main, Yan Yean ...	3	5	0	per cwt.
175.	Ballarat ...	Painting Supreme Court House ...	100	0	0	Completed.
177.	Geelong to Mt. Gellibrand	Repairs to Telegraph Line ...	147	0	0	Completed.
178.	Melbourne ...	Fencing, &c., Yarra Park ...	1,737	8	0	
181.	St. Kilda ...	Extension of Jetty	1,430	0	0	
182.	Mordialloc ...	Extension of Jetty	1,115	0	0	
185.	Geelong, Ballarat	Repairing Telegraph Line ...	193	0	0	
186.	Sandhurst ...	Cooper's Shop, Alterations, &c., Powder Magazine	246	18	6	Completed.
188.	Alexandra ...	Powder Magazine	1,042	0	0	
189.	... ...	Square Telegraph Poles	481	8	0	
191.	Eaglehawk ...	Post and Telegraph Office ...	1,429	18	0	
192.	Rye	Extension of Jetty	616	16	0	
193.	Melbourne ...	Gravel to Victoria Barracks	173	19	0	Completed.
195.	Scarsdale to Rockwood	Telegraph Line	599	0	0	
196.	Sunbury ...	Fencing at Industrial Schools	138	0	0	Completed.
197.	Fitzroy Gardens	Rotunda	275	0	0	Completed.
200.	Kilmore ...	Works and Repairs to Gaol ...	242	5	0	Completed.
202.	Dimboola ...	Police Quarters	298	10	0	Completed.
205.	Melbourne ...	Works at Gaol	133	9	0	Completed.
207.	Buninyong ...	Post and Telegraph Office	1,181	10	10	
208.	Melbourne ...	Works at Exhibition Building for Levee...	127	9	0	Completed.
210.	Toorak ...	Painter's Work, &c. ...	270	0	0	Completed.
212.	Toorak ...	Repairs, &c., to Furniture ...	168	10	0	Completed.
214.	Melbourne ...	Works at Museum, University ...	250	0	0	
221.	Melbourne ...	Fittings and Furniture, Library, Houses of Parliament	300	0	0	Completed.
223.	Ballan ...	Court House	738	17	0	
224.	Melbourne ...	Shed, &c., at Mint	139	0	0	Completed.
226.	Portland ...	Works at Post and Telegraph Office ...	124	0	0	Completed.
228.	Kew Asylum ...	Lodge, Gates, &c.	2,795	0	0	
229.	Macedon ...	Cottages, State Nursery	628	16	6	

			£	s.	d.	
238.	Sandridge ...	Repairs and Extension of Jetty ...	3,320	10	0	
244.	Melbourne ...	Drainage, Public Offices and Parliament Houses	998	18	6	
247.	Melbourne ...	Works at Observatory Buildings ...	187	0	0	
252.	Portland, Chesterton	Repairs, Telegraph Line	491	6	0	
253.	Queenscliff ...	Lighthouse Store	320	0	0	
255.	Maryborough, Talbot, Avoca	Repairing Truck, &c., Powder Magazine	332	0	0	Completed.
256.	Richmond ...	Removal, &c., Lock-up	382	0	0	Completed.
257.	Dunolly and Tarnagulla	Powder Magazine	153	0	0	
258.	Chiltern ...	Lock-up	458	19	0	
267.	Royal Park ...	Powder Magazine Keeper's Quarters ...	483	17	0	
268.	Williamstown ...	Custom House	6,479	14	4	
272.	Casterton ...	Post and Telegraph Office, Court House ...	2,881	0	0	
274.	Melbourne ...	Fencing Government House Reserve ...	616	14	0	
278.	Melbourne ...	Repairs to Roofs, Government Printing Office	172	5	6	
279.	Melbourne ...	Additions, Custom House	26,620	16	4	
282.	Yarra Bend Asylum	Smith and Ironfounder's Work ...	934	0	0	
283.	Mortlake ...	Additions to Post and Telegraph Office ...	567	0	0	
284.	Brighton ...	Additions to Post and Telegraph Office ...	320	16	0	
285.	Melbourne ...	Alterations, &c., Fittings, General Post Office	1,080	12	0	
286.	Jamieson, Alexandra	Telegraph Line	1,110	8	0	
287.	Melbourne ...	Works at Government Printing Office ...	929	0	0	
290.	Eaglehawk ...	Shifting Room, Powder Magazine ...	115	0	0	
294.	Geelong ...	Works at Jetties	2,785	0	0	
297.	Melbourne ...	Fencing, &c., Yarra Park	147	16	0	
299.	Geelong ...	Works at Post and Telegraph Office, &c.	137	10	0	
304.	Melbourne ...	Coal Vault, Treasury...	197	16	6	
305.	Queenscliff ...	Works at Post and Telegraph Office ...	188	0	0	
307.	Colac ...	Works at Post and Telegraph Office ...	117	0	0	
309.	Williamstown ...	Spars for Alfred Graving Dock ...	2,276	9	2	
312.	Nepean ...	Screens, Ocean Park	412	10	0	
314.	Warrnambool ...	Works at Jetty	2,117	7	0	
317.	Melbourne ...	Parade Ground, Russell-street Barracks ...	209	9	0	
318.	Ballarat ...	New Wing to Industrial Schools ...	1,955	7	8	
319.	Maryborough ...	Timber for Gaol Reserve Fencing ...	163	0	0	
320.	Beechworth ...	Works at Court House	216	0	0	
	Wangaratta ...	Post and Telegraph Office ...	2,687	0	0	
	Gippsland ...	Lake Entrance	5,771	18	0	
	Williamstown ...	Keel-blocks for Alfred Graving Dock ...	519	0	0	

P.S.—In the year 1872-3, 138 contracts over £100.
                      210   do.   under £100.

The remarkable list of projects undertaken by the Victorian Public Works Department, supervised by Wardell. The list submitted by him as evidence at the Royal Commission into his Department, June 1873. (This is referred to in the text, and in context, is of particular significance).

BUILDING WITH CONVICTION

APPENDIX 3

Glossary of architectural terms with special reference to their use in the text

Aisle: That part of the church that lies on either or both side of the nave or the chancery.

Ambulatory: The walk or promenade around the apse, behind the High Altar.

Apse: The east end of the church. Vaulted and usually semi-circular lying behind the Chancel, The east end if the orientation is east west. At the north end in St Mary's Cathedral.

Arcade: A line of arches in a gallery, or lining a wall, often 'blind', that is, flanked by the wall itself.

Belvedere, and belvedere tower: The word means, literally, 'beautiful sight' (Italian). It describes a partially open gazebo or lantern usually at the top of a tower affording a fine view of the surroundings.

Boss: An ornamental projection, usually round, not exclusively but most usually placed at the intersection of ribs.

Buttress: A mass of masonry, built against a wall to give it additional strength. Can be internal, but more usually external to the building.

Chancel: That part of the church, usually lying at the east end, beyond the nave and crossing, in which is found the main altar, the choir and the clergy.

Chevet Chapels: A French term to describe the chapels which radiate from the east end of St Patrick's, The term is employed here because Wardell has clearly been influenced by French cathedral design.

Clerestory: Literally, the 'clear story' the upper story of the church (above the Triforium in St Mary's), usually lit by traceried windows.

Coffered ceilings: Sunken panels divided by beams. Often, as in the case of the Gothic Bank in Melbourne, richly decorated.

Columns: Perpendicular pillars, usually circular, supporting arches or Gothic architecture, or entablatures in Classical work. They can be moulded or 'fluted' and stand on plinths.

Corbel: A projecting stone, or sometimes wooden block, which supports a beam or other roof member.

Crockett: A decorative, carved feature projecting from spires or pinnacles.

Crossing: The space where the nave and the transepts meet in a church. In both St Mary's and St Patrick's, the crossing is surmounted by the main towers.

Curvilinear: As distinct from 'Geometric' (q.v.), tracery that is curved and flowing, usually more ornate that geometric., Dates from the beginning of 14th century

Finial: An ornament at the very top of a gable or pinnacle.

Foil: A leaf-shaped formation in most window tracery or a gable, pinnacle, etc. The number of foils is indicated by a pre-fix: trefoil, or quartrefoil.

Flying Buttress: A buttress arch linking the main buttress with the upper wall or vault of the cathedral or church thus transmitting the outward thrust of the forces exerted by the roof.

Gable: The end wall of a building with a pitched roof, triangular, and usually with straight sides. Can have variations – stepped as Wardell employs in some of his banks and warehouses.

Gargoyles: The grotesque carved heads which served as water spouts jutting from roofs. In mediaeval churches they often represent devils ejected from the inside of the church.

Geometric tracery: Differing from curvilinear (q.v.) in its use of circles and simpler geometric forms. Developed in England from the mid-13th Century and favoured by Wardell in his later work.

Groins: The edging stones used where two surfaces meet, as, for example in the vaulting. In Wardell's churches very often he chooses stones of different composition to the main structure.

Hammerbeams: A horizontal roof bracket projecting from the top of the inner wall to carry the arch braces of the timber vaulting. In several of Wardell's churches the hammerbeams also carry painted, carved angels, traditional in mediaeval churches.

Hood-mould: A projecting stone on a wall above a window or doorway designed to throw off the rain

Lights: The glass or openings between the mullions of a window.

Mullions: A vertical stone pillar or division which divides the window into two or more lights.

Nave: The main and usually longest section of the church leading from the west door to the Crossing.

Parapet: A low wall or barrier on or around the top of a building as a protection and ornamentation. It can be battlemented, or with open-work as is the case in Melbourne's Government House.

Pediment: A projecting stone feature above doors and windows in Classical architecture, often gable-shaped.

Piano nobile: Literally, the 'noble floor' (Italian). The principal floor of a building housing reception and ceremonial apartments. In buildings of Italianate or classical style it is usually found on the first floor.

Pierre perdue: In French, literally 'lost stone' – which describes waste rocks and rubble rather than especially shaped and finished stone-work.

Pinnacle: An ornamental pyramid or cone terminating a buttress (usually), or a spire.

Portico: From the Latin, porta, for door or entrance. A projecting, roofed but otherwise open entrance space, often supported by columns as in the case of Government House, Melbourne. May also refer to a recessed entrance.

Rood, and Rood Screen: Rood means Cross or Crucifix and usually refers to the large Cross supported on a beam - or in the case of mediaeval churches, on an ornate screen - dividing the nave from the crossing. Pugin argued for their retention in his churches on historical grounds, and Wardell favoured a simplified form.

Spandrels: The triangular spaces between the sides of arches, their verticals and the horizontals above.

Springing: The point at which an arch springs upward from its supporting columns.

Tracery: The most common feature of all Gothic architecture. The ornamental divisions at the upper section of a window, screen or panel. Often used in blind arches and vaults.

Transept: The transverse arms of a cross-shaped church between the nave and the chancel.

Triforium: An arcaded passage running the length of the nave on either side at a level above the aisle vaults. St Mary's Cathedral has a triforium but is absent in St Patrick's

Vault: An arched roof either in stone or timber.

NOTES AND SOURCES

*Unprinted sources are given in full including the
number of the manuscript and the folio number.*

*For more detailed information on
printed sources, refer to the Bibliography.*

MAHC: *Melbourne Archdiocesan Historical Commission Archives.*
ADAS: *Archdiocesan Archives, Sydney*
BL: *British Library*
LMLA: *London Metropolitan Library Archives*
ML: *Mitchell Library, Sydney*
NA: *National Archives (U.K.)*
SJCA: *St John' College Archives, Camperdown, NSW*
THLHA: *Tower Hamlets Local History Archive*
ADB: *Australian Dictionary of Biography*

INTRODUCTION

1. Confidential Letters to August Lewald on the French Stage, Letter., Trans. by C. G. Leland. Heinman 1893.
2. Manning to Wardell, June 7, 1858. ML. MSS 3728/10
3. Roy Jenkins, *Churchill*, London 2001, p.49
4. Pugin, *Apology.*
5. Paul Johnson, *Cathedrals of England, Scotland and Wales*, p. 65
6. John Betjeman, *Betjeman's Cornwall*, John Murray, London, p.62
7. op. cit., p.52
8. Pugin, Lecture at Oscott College, quoted in Trappes-Lomax, p.153
9. Clark, p.8
10. Wardell entry, *Australian Dictionary of Biography*, Vol. 6, pp.354-355
11. ibid.
12. G. K. Chesterton, 'In La Place de La Bastille' *Tremendous Trifles*, 12th ed. London 1930, p.45
13. For those interested in studying the works of these architects, Brian Andrews' *Australian Gothic* is highly recommended.
14. John Ruskin, *The Seven Lamps of Architecture*, p.190

ENGLAND 1823-1858

1. Birth and Early Life:

1. Ackroyd, *London, The Biography*, p.677.
2. The name 'Tower Hamlets' covers over thirty old districts, one of which included the Tower of London.
3. Masefield, 'Cargoes', *Collected Poems*, London, 1924, p.56
4. Wilson, p. 24
5. For the history of the *Edinburgh Review*, see Hesketh Pearson's *The Smith of Smiths*, from which these facts are taken. Folio Society, London 1977
6. Picard, *Victorian London: the Life of a City*, p.80
7. *Weekly Register*, 'A Few Remarks on Gothic Ecclesiastical Building, and its cost, compared with the Roman and Nineteenth Century Styles',12 January, 1850, p.393
8. De Jong, *William Wilkinson Wardell: His Life and Work: 1823-1899*, Melbourne, 1983, p.9
9. Trustees Minutes, 11/11/1830, p.1064, THLHA
10. Minutes, 23/12/1830, p.1088
11. Minutes, op.cit.
12. *Oliver Twist*, New Oxford edition, p.12
13. Minutes, op.cit., p.1088
14. *Our Mutual Friend*, New Oxford edition, p.509
15. Trustees' Minutes, op. cit., p.521
16. Minutes, op.cit.
17. Minutes, op.cit.
18. Confirmed by Tom Hazell in recorded interview, 5/2/08

2. From Childhood to Maturity

1. *The Times*, 17/10/1834
2. ibid.
3. ibid.
4. Turner's painting of the fire (one of two) is in the Cleveland Museum of Art, USA
5. Wilson, p.10
6. Clark, p.108
7. Both Clark and Trappes-Lomax in their earlier works acknowledge Pugin's contribution but Rosemary Hill's most recent biography most fully documents his work in the new Houses of Parliament, including the design for the tower which is popularly but inaccurately known as 'Big Ben'.
8. V. A. Wardell, 'A Review of the Architecture & Engineering Works of W. W.Wardell', Family Papers, ML.
9. Quoted in Wilson, p.155
10. cf. The 1832 Cholera Epidemic in East London, published in *East London Record*, No.2, 1979
11. V. A. Wardell, op. cit.

12. cf. Family Papers, and letter from Clarkson Stanfield naming Wardell as his 'fellow artist', MSS3728, ML
13. V. A. Wardell, op. cit.
14. Teresa Wardell in conversation with Tom Hazell
15. V. A. Wardell, op.cit.
16. M. Egan, *Journal of the Pugin Society*, Vol.3, No.5, p.57
17. Hill, p.135
18. Colvin, p.31
19. V. A. Wardell, op. cit.

3. *Inveni Quod Quaesivi*
1. Quoted in Holt, p.23
2. Two of the most popular tourist attractions, Westminster Abbey in London, and Notre Dame in Paris, had been gutted of statuary, and their west fronts mutilated. Much of what we see and admire today is nineteenth century restoration work by Sir George Gilbert Scott (Westminster) and Violet-le-Duc (Notre Dame).
3. Hill, p.53
4. Pugin, *The Tablet*, Sept. 25 1852, quoted in Trappes-Lomax, p.57
5. *Weekly Register*, op.cit., p.391
6. Evans, *The Conscious Stone,* passim.
7. Wardell's admiration for Lingard's histories was testified by Teresa Wardell in conversation with her friend, Tom Hazell.
8. Ward, op.cit.
9. Newman to Wardell, 17/7/1853, SJCA
10. *The Times.* op. cit.
11. Newman to Wardell, op. cit.
12. Queen Victoria, *Letters*, ed. Benson, II, p.281
13. It is doubtful whether Wardell's coat of arms was granted by the London College of Heraldry. The surviving coloured illustration preserved in the Wardell Papers, Mitchell Library, although talented, shows a certain casualness and inaccuracy.

4. *The Pugin Connection*
1. De Jong, *Centenary Papers*, p.2.
2. ibid.
3. V. A. Wardell, Memoir , *op.cit.*
4. Quoted in Hill, p.410
5. op. cit., p.315
6. Pugin's churches in Australia include St Francis Xavier's, Berrima; St Charles Borremeo, Ryde, and St Benedict's, Broadway. There are also several examples in Tasmania.
7. Pevsner in the 'Introduction' to Stanton's *Pugin*, p.9.
8. Clarke, p.138
9. Trappes-Lomax, p.258
10. Clarke, p.129

11. Hill, p.93
12. op. cit., p.441
13. Newman quoted in Hill, p.359-360
14. Teresa Wardell quoted by Tom Hazell in interview recorded 5/2/08
15. Hill, p.170
16. *Weekly Register*, op. cit.
17. The extent of the Earl's fortune, and the amount given to the Church has been contested by Denis Gwynn. Although Gwynn does not deny Shrewsbury's extraordinary generosity, he doubts the extent of his fortune and therefore his capacity to give as much as is popularly believed. . See Gwynn, p. xi.
18. Newman, *Letters and Diaries*, Vol.X111, p.460.
19. op. cit., Vol.11, 227.
20. Letter to the *Tablet*, 15 March 1851.
21. Pugin, *True Principles*, p.1
22. op. cit., p.2
23. Wardell to Bishop of Perth, 30 Aug. 1878, Wardell Papers, ML
24. Pugin. Lecture at Oscott College, op. cit.
25. Hill. Letter to author, 19/12/2007
26. ibid,
27. *Weekly Register*, 12/1/1850, p.389
28. Hill, p.486
29. Quoted in Clarke, pp. 143-4.
30. Johnson, *Creators*, p. 148

5. *Architect in the City*

1. There is no record of Littlewood in the RIBA Library. He may have been a non-architect 'sleeping partner' enabling William to set up his practice, or even a phantom figure, added to give the practice additional prestige.
2. I am indebted to Michael Egan for allowing me to make use of his research on this building.
3. Evinson
4. V. A. Wardell, op. cit.
5. Pevsner, *The Buildings of England*, London, Vol.2
6. De Jong, *William Wilkinson Wardell, His Life and Work 1823-1899*, Monash University, 1983
7. Evinson,
8. V. A. Wardell, op. cit.
9. *The Tablet*, 21/10/1848
10. ibid. 28/10/1848
11. ibid.
12. ibid.
13. ibid.
14. ibid.
15. Ladye with an e is the Gothic, or mediaeval form of the word, favoured by both

Pugin and Wardell, and is the accurate title of the Greenwich church.
16. Stebbing, p.24
17. A similar claim is also made for Pugin's church, St Alban's, in Macclesfield.
18. Pevsner, *The Buildings of England*, London, Vol.2: South.
19. Hill, p.410
20. Evinson, p.120
21. Interview with author, 14/10/2007
23. Egan.
24. ibid,
25. Hill, p.482 , and passim.
26. *Tablet*, 28/10/1848
27. ibid.
28. ibid.
29. Pugin writing of Jane Knills, quoted in Hill, p. 407
30. *Weekly Register*, Jan. 1850
31. op. cit.
32. ibid.
33. Newman to Phillips, quoted in Hill, pp389-390
34. *Weekly Register*, op. cit.
35. 'For his mercy is confirmed upon us; and the truth of the Lord remaineth forever.' Psalm 116, familiar to those who still attend the service of Benediction in Latin.
36. Rock to Wardell, 7/1/1856, Wardell papers, SJCA
37. Evinson, *Catholic Churches of London*, op. cit. p.120

6. *The Hampstead Years*
1. Census, 1851
2. Much of this information and the following, given by Pieter van de Merve, biographer of Clarkson Stanfield, to whom I am much indebted for the use of his notes and newspaper cuttings relating to The Green Hill.
3. ibid.
4. cf. *Hampstead Record*, 27/12/1890
5. Stanfield to Wardell, ML - MSS 3728/4
6. Testimonials among Wardell Papers, ML, op.cit.
7. Printed cutting among Wardell family papers, op. cit. undated, but probably mid 1867.
8. Quoted in *The Hampstead Record*, op. cit.
9. Pieter van der Merwe in DNB, Vol.52, (2004 edition), p.99, OUP
10. John Ruskin, *Modern Painters*, Vol. 3, pp.239-40
11. ibid., p.348
12. Stanfield to Wardell, 13/5/48, ML – MSS 3728/4
13. Hilton, p.121
14. Keeley to Lucy Wardell, 7/9/1856, ML - MSS 3728/12
15. Ackroyd, *Dickens*, p. 743
16. ibid.

17. Catherine Dickens to Lucy Wardell, 2/6/1855, ML - MSS 3728/12
18. ibid.
19. Ackroyd, *op.cit.*
20. Wardell family papers, ML - MSS 3728/4.

7.   *Emerging from Pugin's Shadow*
1. Hill, p.344
2. ibid
3. ibid p.364
4. ibid p.361
5. ibid p.486
6. ibid.
7. Hitchcock, *Early Victorian Architecture in Britain,* Vol. 1, p.94
8. *Illustrated London News,* 20/9/1851
9. Tablet,
10. Pugin, *True Principles,* p.1
11. Clark, p.142
12. Wardell to Bennett, 1/7/1854. SJCA.
13. op. cit. 17/7/1854
14. *A History of St Mary & St Michael's Parish,* p. 58, Terry Marsh Publishing, London, 2007
15. ibid., pp.59-60
16. ibid., p.75
17. ibid., p.61
18. ibid., p.65
19. ibid., p.61
20. ibid., p.71
21. de Hels to Superior of the Redemptorists, 7/7/1851, SJCA
22. A copy of the citation is held in the Wardell Papers, ML, MSS3728/4
23. Stanfield to Wardell, ML. MSS3728/1
24. Letter from Charlotte Bronte to her father Patrick Bronte, 7/6/1851. Quoted in Juliet Barker, *The Brontes: A Life in Letters,* p.324. London 1997.
25. *Illustrated London News,* op.cit.

8.   *A Change of Climate*
1. Thomas Hardy to Mary Hardy, 1862, quoted by Claire Tomalin in her biography of Hardy, London 2008. Hardy, at that time met and admired Bejamin Ferrey who worked with Pugin and wrote the first biography of him.
2. The stethoscope was invented by the French physician, Theophile Laennae about 1810 but was not in general use in England until much later in the century. It is possible, though highly questionable whether it was yet widely available in Britain at the time of the diagnosis of Wardell's illness. In George Eliot's novel, *Middlemarch,* the young doctor, Lydgate, just back from medical studies in Paris, uses the 'new-fangled foreign toy' causing scepticism in the community. See

chapter 30. George Eliot's research is considered exemplary. The novel was written in the late 1860s.

3. Roy Porter.
4. *Charlotte Bronte, A Life in Letters,* op.cit. pp.213-4.
5. *Tablet,* 26/6/58
6. Colvin, Howard, 'The Architecture Profession', *Biog. Dictionary of British Architects, 1600-1840,* London 1978
7. Hope-Scott to Wardell, 6/2/1850, SJCA
8. Hope-Scott to Newman, quoted in *Our Lady & St Andrew's: A Parish History,* Feb. 2003.
9. John Dunleavy, 'Pomp and Circumstance: the opening of St Edmund's', *Christian Focus,* Abingdon, Winter, 2007. pp. 6-7
10. ibid.
11. ibid.
12. Stanfield to Wardell, Testimonial, Wardell papers, op. cit.
13. *Tablet,* op.cit
14. ibid.
15. *Weekly Register,* 12/6/1858

MELBOURNE: 1858-1878

9. *The New Settler*
1. Twopeny, p.2
2. Trollope, Vol. 11, p.383
3. ibid.
4. ibid.
5. Trollope, Vol.11, p.389
6. Freeland, *Early Melbourne Churches,*
7. ibid.
8. ibid.
9. ibid.
10. Freeland, *Architecture in Australia*, op.cit, p.102
11. Fitzpatrick to Goold, 15/9/1858, MAHC
12. Wardell to College Council, 6/7/1849, SJCA
13. Fitzpatrick to Goold, 15/11/1858, MAHC
14. ibid.
15. Boland, p.16

10. *'What you do now, do well'*
1. Polding, Pastoral Letter; cf O'Farrell, *Documents in Australian Catholic History,* Vol. 1, p.203
2. Wardell to Connell, 29/11/1858, SJCA
3. ibid.
4. Council Minutes, 24/1/1859. SJCA

5. Wardell to Council, 22/1/1859, SJCA
6. Wardell to Council, 9/5/1859, SJCA
7. Pugin, *True Principles of Pointed or Christian Architecture*, pp.43-46
8. Gorman to Wardell, 18/6/1859, SJCA
9. Wardell to Gorman, 6/7/1859, SJCA
10. ibid.
11. ibid.

11. 'A Quarrel of the Bitterest Kind'
1. *The Australian Builder and Railway Chronicle*, Melbourne, Jan. 1859
2. Amos, p.77.
3. *Age,* 28/10/1864
4, *Age,* 31/1/1859
5. *Age,* 3/3/1859
6. *Age,* 11/4/1859
7. *Age,* 18/10/1852
8. *Age,* 21/10/1852
9. *Age,* 24/10/1852
10. Roberts to Wardell, 17/2/1859, ML. MSS
11. De Jong, *William Wardell: His Life and Work: 1823-1899*, p.41, Monash University, 1983
12. Andrews, p.82

12. A Conflict of Interests
1. Wardell to Council, 30/10/1859, SJCA
2. Johnson memorandum, 12/2/61, Wardell papers. ML
3. Teresa Wardell, in conversation with Tom Hazell.
4. Wardell to Gorman, 23/12/1859, SJCA
5. Gorman to Wardell, 1/12/1859, SJCA
6. Wardell to Gorman, 2/12/1859, SJCA
7. Wardell to Gorman, 7/1/1860, SJCA
8. Wardell to Gorman, 8/2/1860, SJCA
9. Wardell to Eyre Ellis, 31/3/1860, SJCA
10. Wardell to Ellis, 16/5/1860, SJCA
11. Wardell to Ellis, 16/5/1860, SJCA
12. ibid.
13. Wardell to Ellis, 18/6/1860, SJCA
14. Council Minutes, 18/6/1860, SJCA
15. Council Minutes, 18/7/1860, SJCA
16. ibid.
17. *Freeman's Journal,* 29/8/1860, p 2.
18. Wardell to Council, 17/8/1860, SJCA
19. ibid
20. Wardell to Council, 13/5/1860, SJCA

13. *'Grand in its Design'*

1. *Yorick Club Reminiscences May 1868-December 1910*, Melbourne 1911
2. A view supported by one of the members, historian Tom Hazell, in interview with author.
3. David Roberts to Wardell, 17/2/1859, MSS 3728/4, ML
4. Freeland, *Melbourne Churches 1836-1851*, pp128-129
5. Boyd, *The Walls Around Us*, quoted in Andrews, p.81
6. Roberts to Wardell, December 1877, MSS3728/4, ML
7. ibid.
8. Jennifer Wardell, 'Willie's Desk', Wardell Centenary Papers edited by De Jong.
9. Ms letter from Wardell to *Australia Sketcher*, June 1880, ML, original paper not sighted.
10. ibid.
11. Quoted in Freeland, *Melbourne Churches*, op.cit.
12. ibid.
13. ibid.
14. Bolan, illustration on pp.18-19
15. Carr to Wardell, 29/8/94
16. Wardell to Carr, 6/9/94
17. ibid.
18. Boyd, *The Walls Around Us*, op. cit.
19. *Advocate*, 1897
20. ibid.

14. *Under Doctor's Orders*

1. Hamilton's report is held in Wardell Papers, MSS3728/2, ML
2. ibid.
3. Illuminated Address, MS3728/3, ML
4. Edward Wilson to Wardell, 7/7/1870, MSS3728/4, ML
5. ibid.

15. *Management Under Fire*

1. *Age,*
2. Report of Royal Commission, 29/8/73. p. Latrobe Library, Melbourne
3. Wardell memorandum, 12/7/1859, ML
4. Royal Commission Report, op.cit.
5. ibid.
6. ibid.
7. ibid.
8. ibid.
9. ibid.
10. ibid,
11. ibid.
12. ibid.

13. ibid
14. ibid.
15. Freeman, *Architecture in Australia*, p.128
16. ibid.
17. Royal Commission Report, op.cit.
18. ibid.
19. ibid.
20. ibid.
21. ibid.
22. ibid.
23. *Wardell-cum-Eaton*, Report of the Board appointed to enquire into the charges preferred by Mr Eaton ... against Mr Wardell. Oct-Nov 1874. La Trobe Library, Parliamentary Papers.
24. ibid.
25. ibid.
26. ibid.
27. ibid.
28. ibid.
29. *Advocate*, 21/10/1874
30. Wardell-cum-Eaton, op.cit.

16. A Great Volume of Work
1. Trollope, p. 397
2. The present colour scheme dates from 1900. Originally the colour was pearly-grey stucco with bluestone trims in the belvedere.
3. Hitchcock in his *Architecture: Nineteenth and Twentieth Centuries*, gives as a firm opinion that the building 'was consciously modelled on Queen Victoria's Osborne House.' p.171.
4. The likeness is more akin to the Palazzo Senatorio on Capital Hill, Rome. The striking belvedere tower rises up above the balustrade and presents a remarkably similar facade to Government House. Wardell visited Rome on his European tour in 1870.
5. I am indebted to Tom Hazell, one-time officer in residence at Government House, my guide and whose intimate knowledge of the building informs the basis of this section.
6. Bowen to Secretary of State for the Colonies, 4/10/1876, quoted in *Government House Melbourne*, Thomas Hazell, 1994. p.6
7. Quoted in J. K. Ewers, *The Western Gateway*, WA, 1971, p. 244
8. Wardell, 'Report on Fremantle Harbor, Western Australia, and Proposed Improvements.' 1875, p.7
9. ibid.
10. ibid.
11. ibid. See also *West Australian*, 2/2/1875
12. *Argus*,

13. V. A. Wardell, *Memoir*, op. cit.
14. *St Ignatius' Church, Richmond*. Guide book, 2004, p.7
15. Teresa Wardell in conversation with Tom Hazell
16. ibid.
17. V. A. Wardell, op.cit.

17. *Black Wednesday and Departure*

1. Garden, p.149
2. ibid., p.50
3. *Argus*, 9/1/78
4. ibid.
5. *Age*, 12/1/78
6. Vincent Wardell, op. cit.
7. *Cyclopaedia of Victoria*, p. 102
8. ADB, 1851-1890, p.205.
9. *Advocate*, 1878

SYDNEY 1878-1899

18. *'New South Wales Enriched'.*

1. S Matthew's Gospel, 10 v 14-23, RSV edition
2   *Tablet*, 27/4/78
3. *Sydney Morning Herald*,
4. Wardell to Finance Committee, *30/1/78*, SAA
5. ibid.
6. Author's conversation recorded with Tom Hazell, old friend of Wardell family, op.cit.
7. Wardell to Finance Committee, ADAS
8. Vaughan to Fr. E. Guy, quoted in *St Mary's Cathedral, 1821-1971*, p.130
9. *Sydney Morning Herald*, 17/9/78 quoted in *St Mary's Cathedral*, 1821-1971, *op.cit.*
10. A. E. Cahill, 'Archbishop Vaughan and Cardinal Moran as Cathedral Builders,' in *St Mary's Cathedral 1821-1971*, pp.134-5
11. Wardell to Vaughan, 16/3/82, ADAS
12. [not only] 'for the use of the people' but 'to beautify the city'
13. Wardell to Vaughan, 30/8/78 ML, MS10/1,
14. Wardell to Parry, 30/8/1878.
15. ibid.
16. ibid
17. Jones, Michael, *North Sydney, 1788-1988*, pp.96-97, Sydney 1988
18. ibid
19. Wardell to Verdon, Wardell Letter Books, ML
20. Wardell to Fitzpatrick, 13/9/78

19. *A Phoenix from the Ashes*

1. 'St Mary's Cathedral, Sydney, A Memoir of Its Destruction by Fire, 1865,' quoted in *Benedictine Pioneers of Australia*, Vol 11, pp.289-295
2. ibid.
3. ibid.
4. ibid.
5. *St Mary's Cathedral, Sydney,* 1821-1971, op.cit. , Document 9, pp.90-97
6. ibid.
7. ibid.
8. ibid.
9. Polding to Wardell, 10/10/1865, ADAS
10. ibid.
11. Teresa Wardell in conversation with historian Tom Hazell, op.cit.
12. Wardell to Polding, 20/10/1865, ADAS
13. ibid.
14. *Benedictine Pioneers*, op.cit.
15. Donovan to Hardman, 9/9/1882, ADAS
16. ibid.
17. 4/5/82
18 16/9/86
19. The recent addition of wooden screens enclosing the sanctuary and choir restrict the glorious vista previously possible from the ambulatory across the chancel, and negate the original function of Wardell's railings
20. Freeland, *History*, p.127
21 McDonald, ADB, Vol. 6, pp.354-355

20. *Architect of Commerce and Industry*

1. St Mary's Cathedral, Sydney, 1821-1971, p.137
2. Freeland, *History*, p.125
3. I am indebted here to Ursula De Jong for information in her *Exhibition Catalogue, William Wardell: His Life and Work*, p.61
4. *Sydney Morning Herald*, 5/3/87
5. Wardell to Union Club, 19/5/85, ML
6. ibid.
7. Wardell to Mahoney, 1/2/86, ML
8. ibid.
9. ibid.
10. *The English, Scottish and Australasian Bank*, ML, MSS 3728/12.
11. Sir George Verdon entry in ADB, Vol. 9, p.330-2
12. *Illustrated Australian News*, 3/10/83
13. Boyd, *Australian Ugliness*, Melbourne 1960, p.63
14. 'The Gothic Bank' was voted by readers as Melbourne's favourite building in *The Age*, 1989.
15. Lea-Scarlett, *Riverview: A History*, Hale & Iremonger, 1988.

16. Dalton Diary, facsimile edition held in Ignatius College archives, Riverview.
17. Wardell, Report on Calliope Dock to Chairman, Auckland Harbour Board, 27/7/83
18. cf. Evans, *C. Y. Connor, His Life and Legacy*, Ch. 15.

21. A Position of Eminence
1. Wardell, 1/2/86. ML
2. Wardell to Vaughan, 3/7/81. ADAS
3. Wardell to Bell, 19/7/81
4. ibid,
5. Wardell to Dwyer, 28/1/69. ADAS
6. A. E. Cahill, essay in *St Mary's Cathedral, 1821-1971*, p.137
7. Wardell to Vaughan, 30/3/82. ADAS
8. Undated letter from Wardell to Vaughan on Upton Grange notepaper
9. Wardell to Vaughan, 16/3/82. ADAS
10. Wardell to Vaughan, 25/8/82 ADAS
11. *St Mary's Cathedral 1821-1971*, Document 13, pp.108-9
12. ibid,
13. ibid.
14. ADB, pp.330-332
15. Union Club Secretary to Wardell, ML. 687.
16. Wardell to Union Club, 19/5/85, op.cit.
17. Ayres, p.180
18. ibid., p.272
19. Wardell to Fry, 19/12/88. ML
20. *Sydney Morning Herald report*, 9/12/89
21. Wardell to Barling, 3/1/90. ML
22. Wardell to Fry, 9/1/90
23. Wardell to Building Committee, 17/3/90
24. Wardell to Carr, 18/6/90. ML
25. Wardell to Carr, 27/5/91
26. Wardell to Todd, 18/7/91
27. Wardell to Byrne, 16/10/91
28. Wardell to Hardman, 24/10/91

22. The Final Years
1. *Freeman's Journal*, 23/4/92
2. ibid.
3. ibid.
4. Wardell to Murphy, 14/11/92
5. op. cit., 15/12/92
6. Wardell to Carr, 9/8/93
7. T. A. Coglan, *Encyclopaedia Britanica*, 11th Edition, Vol.2, p.965
8. Campbell to Wardell, 26/11/94

9. Wardell to Campbell, 29/11/94
10. Wardell to St Mary's Building Committee, 22/3/90
11. Glanville, 'Architecture, Building and Architect' in *St Mary's Cathedral Sydney*, op.cit.
12. Wardell to Secretary, Board of Health, Melbourne, 25/1/94
13. Wardell to Carr, 17/6/95
14. Wardell to Farmer and Brindley, 29/2/96
15. Wardell to Building Committee, 10/5/89. ADAS.
16. Wardell to Moran, 1/11/98, ML
17. Wardell to Dean O'Haran, 24/6/99, ML
18. ibid.
19. Wardell to Moran, 15/9/99, ADAS.
20. Wardell to O'Haran, 13/11/99
21. Blessed Mary McKillop's remains and permanent shrine are now located at the Mary McKillop Centre, in North Sydney, around the corner from Wardell's old home Upton Grange in Edward Street.
22. Herbert Wardell to Dean O'Haran, 28/11/99
23. *Sydney Morning Herald*, 22/11/99
24. *Freeman's Journal*, 25/11/99
25. *Tablet*, 6/1/1900
26. *Age*, 21/3/1936
27. Doorly, p.1

BIBLIOGRAPHY

Published works consulted and referred to in **The Notes:**

Ackroyd, Peter, *London: The Biography*, London. 2000
 – *Dickens,* London 1993
Alexander, Michael, *Medievalism: The Middle Ages in Modern England,*
 Yale University Press, 2007
Amos, Keith, *The Fenians in Australia 1865-1880*, NSW University
 Press, 1988
Andrews, Brian, *Australian Gothic, The Gothic Revival in Australian
 Architecture from the 1840s to the 1950s,* The Miegunyah Press,
 University of Melbourne 2001
Anson, Peter, *Fashions in Church Furnishings 1840-1940*, London 1965
Ayres, Philip, *Prince of the Church: Patrick Francis Moran 1830-1911*,
 Melbourne 2007
Bessant, Sir Walter, *Survey of London,* "London in 19th Century",
 London 1909
Boland, T. P. *St Patrick's Cathedral: A Life,* Polding Press, 1997
Boyd, Robin, *The Australian Ugliness,* F. W. Cheshire, Penguin edition,
 1965
Burt, H. N. *Benedictine Pioneers of Australia,* Lond. 1911 Vols 1&ll
Chesterton, G. K. 'In La Place de La Bastille' *Tremendous Trifles,* 12th ed.
 London 1923
Clark, Kenneth, *The Gothic Revival,* 3rd Ed. London. 1974
Crowther, M. A., *The Workhouse System 1834-1929*, Batsford Academic,
 London. 1981
Dawson, Christopher, *The Spirit of the Oxford Movement,* St Austin
 Press reprint, London 2001

De Jong, Ursula M. (Ed.), *The Architect & His Era: Centenary Papers,*
  – DeakinUniversity, Geelong, 2000
  – *William Wilkinson Wardell, 1823-1899,* Monash University, 1983.
  – *St Patrick's Cathedral & William Wardell, Architect,* Catalogue
  – Essay, Archdiocese of Melb. 1997
  – *St Patrick's Cathedral Melbourne: A Guide,* Catholic Archdiocese of
  Melb. 2005
Doorly, Moyra, *No Place For God,* Ignatius Press, 2007
Egan, Michael, 'The Early History and Work of William Wilkinson
  Wardell', in the Journal of the Pugin Society, Vol.3, No. 5
Evans, A. G., *The Conscious Stone,* Polding Press, Melb. 1984
  – *C. Y. O'Connor: His Life and Legacy,* UWA Press, 2001
Evinson, Denis, *Catholic Churches of London,* Sheffield Academic Press,
  1998
Freeland, J. M., *Architecture in Australia: A History.* F.W. Cheshire, 1970
  – *Melbourne Churches 1836-51,* Melbourne,.
Garden, Don, *Victoria: A History,* Nelson, Melb. 1984
Gardner, S, A *Guide to English Gothic Architecture,* Cambridge,
  2nd ed. 1925
Gwynn, Denis, *Lord Shrewsbury, Pugin and The Catholic Revival,*
  London 1946.
Herman, Morton, *The Early Australian Architects & Their Work,*
  Sydney 1954
Hill, Rosemary, *God's Architect: Pugin and the Building of Romantic
  Britain,* London 2007
Hilton, Tim. *John Ruskin,* Yale, 2000.
History of St Mary & St Michael's Parish, Published by the parish in
  association with Terry Marsh Publishing, London 2007
Hitchcock, Henry-Russell, *Architecture: Nineteenth and Twentieth*
  – *Centuries,* Penguin Books, Baltimore. 1958
  – *Early Victorian Architecture in Britain,* New York, 1954
Johnson, Paul, *Cathedrals of England, Scotland and Wales,* London 1993.
  – *Creators: From Chaucer and Durer to Picasso and Disney,* London, 2006
Martin, Christopher, *A Glimpse of Heaven: Catholic Churches in England
  and Wales,* London 2007

Merwe, Pieter van der, *Clarkson Stanfield*, Nat. Dictionary of Biography, Vol.52, pp.98-103, OUP, 1996

Newsome, David, *The Convert Cardinals: Manning and Newman*, London, 1993

– *The Victorian World Picture*, Rutgers University Press, New Jersey, 1997

Newman, John Henry, *Letters and Diaries*, Vols. X11, Xlll and XlV. Ed by C. S. Dessain, 1963

O'Farrell, P., *The Catholic Church in Australia: A Short History 1788-1967*, Melbourne, Reprint ed. 1975. (Ed.)

– *St Mary's Cathedral Sydney 1821-1971*, Devonshire Press, NSW 1971

Picard, Liza, *Victorian London, The Life of a City*, London 2005

Porter, Roy, *The Greatest Benefit to Mankind: A Medical History of Humanity from Antiquity to the Present*, London 1997

Pugin, A.W.N., *The True Principals of Pointed or Christian Architecture*, Reprint of 1st ed. St Martin's Press, New York

– *Contrasts; or a parallel between the noble edifices of the fourteenth and fifteenth centuries, and similar buildings of the present day; showing the decay of taste; Accompanied by appropriate texts*, Lond. 1841.

Rolt, L. T. *Victorian Engineering*, London 1970

Ruskin, J. *The Seven Lamps of Architecture*, Everyman edition (No.207) Lond. 1956

– *The Stones of Venice*, Folio Society ed. 1987

Stanton, Phoebe, *Pugin*, Lond. 1971

Stebbing, George, *History of St Mary's, Clapham*, London 1935

Trappes-Lomax, Michael, *Pugin: A Mediaeval Victorian*, London 1933

Trollope, Anthony, *Australia and New Zealand*, Vols 1 & 1l, 2nd ed., reprint, London 1968

Trevor, Meriol, *Newman: The Pillar of the Cloud*, N.Y. 1962

Ward, Wilfred, *The Life of John Henry Cardinal Newman*, Vol.2, London, 1912

Wilson, A. N., *The Victorians*, Arrow Books, London, 2003

Young, G. M., *Victorian England: Portrait of an Age*, OUP, 1936

Periodicals

Advocate
Age, 1858-1870
Australian Builder, 1885-1925
Register, 1850
Sydney Morning Herald, 1870-1899
Tablet, 1899-90
The Times 1847
Western Australian Times, 1875

Manuscripts and Unpublished Papers

De Jong, Ursula M. 'From England to Australia: the Architecture of William Wilkinson
Wardell (1823-1899).' Doctoral Thesis, Monash University,
Clive Lucas, Stapleton and Partners, St John's College, Conservation Management Plan, 2001

Libraries Consulted:

MAHC, Melbourne *Archdiocesan Historical Commission.*
ADAS, *Archdiocesan Archives, Sydney*
BL, *British Library*
LMLA, *London Metropolitan Library Archives*
ML, *Mitchell Library, Sydney*
NA, *National Archives (U.K.}*
SJCA, *St John' College Archives, Camperdown, NSW*
THLA, *Tower Hamlets Library & Archive*

ACKNOWLEDGEMENTS

A popular belief holds that fiction can be written by a reclusive author alone in his garret - a conclusion likely to be denied by novelists who undertake extensive research for their plots. On the other hand biographies invariably depend on much input, advice and encouragement from experts in related fields, and from the support of friends. None more so than this work on Wardell which could not have been written without the generosity and cooperation of many people. I owe a great debt to the following contributors and the order in which they are placed should not be read as an order of precedence; each one played a vital part in the completion of the work.

In England I was greatly assisted by John Dunleavy who was my guide on various excursions, and to my daughter, Emily, for her valuable research assistance which continued there after my return to Australia. My guide in Scotland was Irene Furlong, native of the Borders, who was most generous with her time and transport. Irene, with her local connections, was able to arrange a private tour of Abbotsford House. In Greenwich, Pieter van der Merwe generously made his researches on Clarkson Stanfield and The Green Hill available to me. Others in England who answered my questions and gave advice included Rosemary Hill, the much-praised biographer of Pugin – a work of which I made great use. I thank also Don Wilkes, a descendent of the Butler family of Oxford, and Michael Egan, Archivist of the Parish of Our Ladye Star of the Sea, who graciously made his research on Wardell and his early work in England available to me; his knowledge of the history of Wardell's Greenwich church is unsurpassed. I received courteous help from Librarians at the Tower Hamlets Local History Archive and Library, and at the London Metropolitan Library, the London Library and the National Archives at Kew. Courtesy was also shown to me by various parish priests who welcomed me, opened their churches and answered my questions.

I also remember with gratitude the nuns of Tyburn Convent who made accommodation available for me in central London, and no doubt supported my work with their prayers.

In Melbourne I was assisted by Rachel Naughton, the archivist of the Archdiocesan Historical Commission; and I owe special thanks to Tom Hazell who was not only my tireless guide and adviser, but who also read the manuscript and made many helpful comments and corrections. I also wish to thank Ursula De Jong, Architectural historian, for her advice and interest in my work; no one can start writing on Wardell without drawing heavily on De Jong's expert studies and her early championing of Wardell's architecture.

In Sydney I was greatly assisted by Pauline Garland, Archivist of the Sydney Archdiocese, and the Librarians at the Mitchell Library. Thanks are due to David Daintree, former Dean of St John's College, who allowed me access to the Wardell papers, and to Archivist, Dr Perry McIntyre, who welcomed me and guided me through the material. I thank Charity Haynes for helping with accommodation in Sydney and especially to Richard and Cynthia Connolly who made their house available for an extended period and cheered me with undeserved praise and encouragement.

My mentor, adviser and constant support has been Father Paul Stenhouse, MSC whose initial encouragement set me on the Wardell road and kept me there even when the going was tough and morale was low. His reading of an early manuscript and his comments were invaluable. The final work owes much to him. My wife Claire, as ever, provided constant support, read my efforts as they came off the printer and admonished me at appropriate times.

Finally, I would like to acknowledge with sincere thanks the generous financial support of His Eminence, Cardinal George Pell, and Archbishop Denis Hart of Melbourne. Without their help - notwithstanding the support of others listed above – the book simply could not have been attempted.

I trust that the result offered here has been in some way worthy of all the help that I have received.

INDEX

(figures in bold type denote illustrations)

Goodford, George, 240
Goold, Bishop James Alipius, 85, 114, his character, 115; 117; his idealism, engages WW, 118; 119, 121, 131
Gore Hill Cemetery, 10
Gorman, John V. 125, 143
Great Exhibition 1851, 72, 80-81
Green Hill, The, (see Hampstead)
Greenway, Francis, 11

Hackney, 16, 61
Hamilton, A. S. 158-159, 163
Hampstead, description of , 61-62; **96, 97**
Hansom, J. A., 55, 73, 184
Hardman, John H., 40, 55, 88, 233
Hardy, Thomas, 83
Harris, Sir Mathew, 237
Harvey, John James, 164
Hawes, John Dean, 35
Heine, Heinrich, 7
Herbert, John Rogers, 42, 47, 66, 149, 160
Hexam, Lizzie, 18
Higden, Betty, 23
Higinbotham, Thomas, 182
Hill, J. F., 142
Hill, Rosemary, 31, 47, 71
Hixson, Francis, 207
Hogarth, Georgina, 68
Hope-Scott, James, 47, 86-88, 184
Hope-Scott, Charlotte, 86
Horne, G. S. W., 129
Hunt, Horbury John, 12, 250
Houses of Parliament, Westminster, the great fire, 27-28; 86
Howard, Thomas, 19

81; marriage, 49-50, 61; character, 45, 129, 145, 225-226; professional standards, 155, 202-203, 219, 234-235; as a family man, 149, 179; his health, 83-84; 202, migration to Australia, 83-85, 90-92; arrival in Melbourne, 113; description of Melbourne, 1858, 113-114; faces sectarian conflict, 131-133, 180; staff, relationships and management, 159-160, 164-165, 170-172; consults phrenologist, 157-159; sick leave and travels in Europe, 159-161; volume of work, 163-164, 215, (see also *Appendix 2*, pp271-273); Royal Commission examines WW's Department, 166-170; investigation into suitable location of Fremantle Harbour, 175-178; dismissal from Public Service, 181-185; leaves Melbourne for Sydney, 201-202; ES&A Banks, his work on: 205-206; submits designs for Perth Cathedral, 206-207; religious convictions, 9; his accident, 243-244; last illness's and death, 249-250; and obituaries, 250-252; his will and legacies, 252-253

His works:

Lightning Source UK Ltd.
Milton Keynes UK
UKHW02f0613060618

323809UK00010B/532/P